Integrated Water Resources Management in South and South-East Asia

Books in the Series

Integrated Water Resources Management in South and South-East Asia

edited by
Asit K. Biswas
Olli Varis
Cecilia Tortajada

OXFORD
UNIVERSITY PRESS

OXFORD
UNIVERSITY PRESS

YMCA Library Building, Jai Singh Road, New Delhi 110 001

Oxford University Press is a department of the University of Oxford. It furthers the
University's objective of excellence in research, scholarship, and education
by publishing worldwide in

Oxford   New York

Auckland   Cape Town   Dar es Salaam   Hong Kong   Karachi   Kuala Lumpur
Madrid   Melbourne   Mexico City   Nairobi   New Delhi   Shanghai   Taipei   Toronto

With offices in

Argentina Austria Brazil Chile Czech Republic France Greece Guatemala
Hungary Italy Japan South Korea Poland Portugal Singapore Switzerland
Thailand Turkey Ukraine Vietnam

Oxford is a registered trademark of Oxford University Press
in the UK and in certain other countries

Published in India by Oxford University Press, New Delhi

© Oxford University Press 2005

ISBN 019 566932 0

Printed in India by Pauls Press, Delhi 110020
Published by Manzar Khan, Oxford University Press
YMCA Library Building, Jai Singh Road, New Delhi 110 001

PREFACE

While forecasting the future is an extremely hazardous business, one thing can be predicted with complete certainty: the world in the year 2024 will be vastly different from what it is today. This trend, of course, is neither new nor unexpected, since the world today is very different from what it was in 1984. The differences between the developments during the next twenty years compared to the preceding twenty years will be that the changes during the coming decades will be significantly faster and more extensive than what have been witnessed in the past. The speed of these developments will make efficient and rational management of these changes a very complex and difficult task under the best of circumstances.

Among the main driving forces that are likely to contribute to these changes are rapidly evolving demographic conditions, simultaneous urbanization and ruralization in nearly all developing countries, accelerated scientific and technological advances in all fields, speed and nature of globalization, communication and information revolution, improvements in human capital, and steadily increasing expectations of the people of the developing world in terms of poverty alleviation and better quality of life.

The water sector is an integral component of the global system, and thus changes in the global system may affect the water sector through a variety of ways, some known but others unknown. To a significant extent, the global system, in turn, is likely to be affected by how the water resources are planned and managed. The developing world would be much more affected in terms of how its water resources are managed compared to the developed world, whose economies are now more resilient in terms of its linkages to water-related issues.

In order to meet the challenges imposed by the rapid global changes, existing water management systems must also change commensurately. Global forces outside the water sector are already shaping the future availability and patterns of use of this resource, and thus its management practices. These impacts are likely to increase very significantly during the next quarter of a century. Thus, professionals associated with the water sector must seriously examine the implications of the changes to planning and management of water.

In view of the rapid changes expected in the coming years, it is essential to examine critically and objectively the current paradigms and theories that are being used for managing the water sector. One of these current popular paradigms is integrated water resources management (IWRM).

The concept of IWRM, however, is not new. Many water professionals have been arguing for the use of an integrated concept from as far back as the 1940s, that is for more than half a century. The concept was strongly recommended during the first and only United Nations Water Conference that was held at a high decision-making level at Mar del Plata, Argentina, in March 1977. Some fifteen years after the Mar del Plata recommendations, IWRM was again recommended during the International Conference on Water and the Environment, held in Dublin in 1992. Agenda 21 of the Rio Conference gave IWRM a further boost. Numerous conferences, symposia, and workshops have been held on IWRM in different parts of the world since 1992, nearly all of which have recommended the concept without any reservation. So also did the World Summit of Sustainable Development in Johannesburg in 2002, as well as the Third World Water Forum in Kyoto in 2003. Consequently, most external support agencies that have an interest in water issues have taken considerable interest in the process of IWRM, both directly and indirectly.

Unquestionably, global paradigms like IWRM are conceptually attractive, but it is at present an open question as to the extent to which it is possible to move from the concept to the implementation phase. It is thus essential to objectively analyse its applicability: conceptual attractiveness alone is no solution to ensure efficient and equitable water management in the future. Furthermore, contrary to what many may believe, articulation, debate, and implementation of paradigms may be the quickest way to ensure efficient management of water resources.

The main objective of this book is to examine the concept of IWRM and its current implementation status in the real world, critically, objectively, and without being dogmatic about it. The issue of IWRM is analysed in its totality with an open mind, without any preconditions, biases for or against, or hidden agendas. The focus is to examine the implementation status and potential of the concept under the present and foreseeable conditions for a specific region of the world. The region selected is South and South-east Asia, where water is an essential requirement for promoting regional development and poverty alleviation, and where water management and development has a long history. This region is also a priority area for development, since according to the World Bank, some 48 per cent of the world's absolute poor (daily income less than US$1) live in South and South-east Asia.

It is thus imperative to identify linkages between IWRM and poverty alleviation so that IWRM could be an efficient instrument to improve the standard of living and quality of life of the absolutely poor people in South and South-east Asia, and to see what have been IWRM's overall impacts on the people and the environment. The current status of the use of the concept of IWRM to improve water management is examined through a series of case studies from various countries of the region. The case studies are neither academic nor theoretical, but rather based on what has happened or not happened in terms of implementing the concept of IWRM in South and South-east Asia, and what have been their benefits and disadvantages.

The focus of each of these case studies is to answer questions like the following:

(i) How has each country, river basin, or a smaller geographical unit, on which the case study is based, has defined IWRM?

(ii) To what extent has the IWRM concept been systematically applied in the geographical unit considered? What have been the institutional and legal arrangements through which IWRM concept was applied?

(iii) What have been the overall results (positive, negative, and/or neutral) on water management because of the implementation or non-implementation of the concept?

(iv) What have been the impacts in terms of poverty alleviation, employment generation, improving quality of life, and environmental conservation?

(v) If the concept has worked, what were the conditions that made it work, and why? What can be done to further improve the operationality and efficiency of the concept?

(vi) If the concept did not work, what were the main reasons? Can these constraints be successfully overcome so that the concept can be implemented? If not, what practical steps should be considered?

(vii) Based on these experiences, what lessons can be learnt? If the concept works, how can its operationalization be made more efficient and widespread? If it does not, should a new paradigm or pluralism of paradigms, be considered, depending upon social, economic, institutional, and other relevant conditions for each specific case, and/or region?

The case studies are from Bangladesh, China, India, Indonesia, Malaysia, Nepal, Thailand, and Vietnam. Also the Mekong River Commission was invited to prepare a paper outlining their experiences with IWRM. Each case study attempts to answer the set of questions mentioned above so that the results should be comparable.

In addition, three papers with a more general scope were commissioned. The first deals with the development processes in South and South-east Asia on a general level, and links water as an integral part of this process. The second one summarizes the experience on IWRM in Latin America and gives a comparative perspective on how issues pertaining to water are managed in that part of the world. The third paper is a critical reassessment of IWRM in operational terms.

The papers were presented and thoroughly discussed in a workshop that was held at the Asian Institute of Technology in Bangkok, Thailand, in December 2002. Many issues that are currently associated with IWRM and are mostly unquestioned were reviewed objectively and critically. Among such issues were:

  (i) Water sector is much more dependent on the rest of the society than most IWRM literature appreciates.

 (ii) IWRM must be tailored and comprehended in a very case-specific way. One size does not fit all.

(iii) Whereas IWRM has been articulated as the best means to water resources management, not a single large project could be found where IWRM has been applied with full success.

(iv) The discussion on IWRM in international forums tends to forget that we should not only talk about management of water resources, but even more importantly, focus on development of water resources in specific regions such as South and South-east Asia. Both development and management should be made in an integrated way—an issue which is often forgotten in countries which already have a highly developed infrastructure and governance in place.

 (v) The river basin may not necessarily be the most relevant management unit. This is due to many issues such as international boundaries, distribution of economic activities, virtual water, urban systems, historical and jurisdictional borders within national boundaries, groundwater aquifers which often do not go along surface water basins, and so forth.

(vi) The governance structure in the analysed countries and regions is in reality very far from the idealistic picture drawn by IWRM rhetoric. Lack of policies or laws may not be the reason for malfunctions but the implementation phase as well as a proper functioning of the governance system in totality is more critical.

It is easy to agree with Cecilia Tortajada when she says: 'In general, an objective review of the situation in the region indicates that in the area of IWRM, there are more questions than answers, and more challenges than achievements.' This is how she assessed the issue in her thorough commentary of the workshop that was published both in *Water International* in March 2003 and in *International Journal of Water Resources Development* in June 2003.

The study was jointly carried out by the Helsinki University of Technology and the Third World Centre for Water Management. The funding came from the Department of International Cooperation, Ministry of Foreign Affairs of Finland, for which we are very grateful.

ASIT K. BISWAS
OLLI VARIS
CECILIA TORTAJADA

CONTENTS

TABLES AND FIGURES

Tables

Figures

1

EXTERNALITIES OF INTEGRATED WATER RESOURCES MANAGEMENT (IWRM) IN SOUTH AND SOUTH-EAST ASIA[1]

Olli Varis

The concept of integrated water resources management (IWRM) is included in almost all contemporary water agendas, irrespective of the purpose and scale of the task. Accordingly, the concept is understood in many ways and a number of definitions exist. Most commonly, the holistic and coordinated management of all the various aspects and uses related to water are emphasized. The most used definition of IWRM these days is the one by the Global Water Partnership (GWP, 2000):

IWRM is a process which promotes the coordinated development and management of water, land, and related resources, in order to maximize the resultant economic and social welfare in an equitable manner without comprising the sustainability of vital ecosystems.

The concept is not new. However, it has become increasingly pronounced during the past one or two decades, obviously as a response to the growingly recognized problems due to too fragmented management of water. Fragmented in a multitude of ways, as can be seen in subsequent chapters.

A few milestones of the concept IWRM are worth mentioning here. The most quoted document from the 1990s in this respect was obviously the resolution of the UN International Conference on Water and Environment that was held in Dublin, Ireland, in 1992 as a main preparatory water event for the Rio Conference. The well-known Dublin Principles are:

[1] This study has been funded by the Ministry of Foreign Affairs of Finland and the Academy of Finland under the project 45809. Special thanks are due to Asit K. Biswas, Mohammed Mizanur Rahaman, Tommi Kajander, Katri Makkonen, and Pertti Vakkilainen for their fruitful cooperation.

- Freshwater is a finite, vulnerable, and essential resource, which should be managed in an integrated manner;
- Water development and management should be based on a participatory approach, involving users, planners, and policy makers at all levels;
- Women play a central role in the provision, management, and safeguarding of water;
- Water has an economic value and should be recognized as an economic good, taking into account affordability and equity criteria.

The first principle clearly articulates the need for IWRM. However, these principles were not influential enough to bring water high on the agenda in Rio. To the contrary, the water community largely shares the view that water did not receive the attention it deserves in the discourse of sustainable development.

Gradually the water profession raised its head in the latter part of the decade. Most visible undertakings include the creation of the World Water Council and the Global Water Partnership, as well as the organization of several major international events around water and development issues, such as the World Water Forums and Stockholm Water Symposia.

As the Dublin Conference prepared a water agenda for Rio, the Bonn Freshwater Conference of December 2001 did the same for the Johannesburg Summit. IWRM was also on agenda, as condensed in the so-called five Bonn Keys:

   (i)   The first key is to meet the water security needs of the poor;

  (ii)   Decentralization is key. The local level is where national policy meets community needs;

 (iii)   The key to better water outreach is new partnerships;

 (iv)   IWRM is needed to bring all water users to the information-sharing and decision-making tables;

  (v)   The essential keys are stronger and better performing governance arrangements.

The Johannesburg Summit took the issue of water as one of its ten focal areas, and thus many think that water is far better taken into consideration than was done in Rio. The Framework for Action of Johannesburg includes many interesting recommendations with respect to water. Some of the most important ones include the following:

   (i)   Developing IWRM and water efficiency plan by 2005 for all major river basins of the world;

  (ii)   Developing and implementing national/regional strategies, plans, and programmes with regard to IWRM;

 (iii)   Improving efficiency of water uses;

(iv) Facilitating public-private partnership;
 (v) Developing gender sensitive policies and programmes;
(vi) Involving stakeholders, specially women, in decision making, management, and implementation processes.

The first two items in the above list are very interesting with respect to IWRM. They define clear operational targets for the implementation of IWRM both in the river basin context as well as on the jurisdictional level. How realistic these targets are is another question, but what is important is that IWRM is very high on the Johannesburg agenda.

Development of water resources is a part of the more general development process. As the subsequent sections comprehensively scrutinize the situation of IWRM in various Asian countries and river basins, this chapter concentrates on providing a systematic analysis of the external setting in which water resources management and development must take place. Largely, this means the many-sided context—which consists of the people, society, economy, and resources—within which water policies are formulated.

This analysis makes a special reference to the results of the South Asia Water Vision by the Global Water Partnership (SASTAC, 2000). Other references are made as well. Of particular interest is whether the Vision takes these externalities appropriately into consideration. Unfortunately, the Vision document for South-east Asia is not concrete enough for this purpose.

South and South-east Asia are two of the five critical regions, which have been chosen as focal regions in the analysis of interconnections between water, food, poverty, and urbanization (Varis, 2000; Vakkilainen and Varis, 1999; Varis and Vakkilainen, 2001). The other regions are China, the Nile Basin, and West Africa. Of the world total, they account for 59 per cent of the total population, 80 per cent of urban population, 60 per cent of urbanization, 7 per cent of GNP, 34 per cent of arable land, 48 per cent of cereal production, and 57 per cent of irrigated land.

Ten critical externalities—or driving forces—were identified as the most critical ones with respect to water resources development in South and South-east Asia (Figure 1.1). These are:
  (i) Population;
 (ii) Urbanization;
(iii) Low and uneven human development;
(iv) Retarded decentralization and controversies in empowerment;
 (v) Slow globalization and regional integration;
(vi) Malfunctions in governance and policy environment;

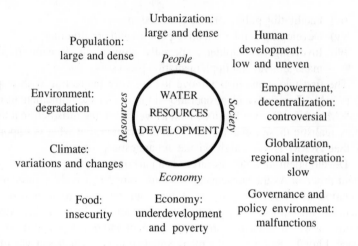

Figure 1.1: Critical externalities of water resources development in South and South-east Asia

(vii) Economic underdevelopment and widespread poverty;
(viii) Food insecurity;
(ix) Variations and changes in climate; and
(x) Environmental degradation.

These will be discussed one by one, and compared with the situation in the regions mentioned above, with the exception of the environmental issues. Water and nature is one of the strongest component of the South Asia Water Vision, and plenty of other profound analyses of the water environment of the subcontinent are available. Therefore, only some remarks are made in the conclusions.

Prior to the discussion of the key externalities, summaries of water resources of South and South-east Asia are presented.

South Asia and its Waters

The combined population of India, Nepal, Pakistan, and Bangladesh approaches 1.4 billion and it is growing fast. In 2025, the population is predicted to approach 1.9 billion. Gross National Product (GNP) per capita in the area is low: US$ 470 per annum in Pakistan, US$ 440 in India, US$ 220 in Nepal, and US$ 370 in Bangladesh.

Over half of India's land area is used by agriculture. Arable land accounts for 188 million ha. It is difficult to increase this area. To provide enough food for the increasing population, agriculture has to be made more effective. This demands even more intensified measures. The irrigated land area is now about 38 per cent of all arable land. It is possible to improve irrigation but new projects are economically demanding and more difficult to implement than before. The use of fertilizers per arable hectare has grown sevenfold during the last twenty-five years.

The total amount of India's renewable water resources according to SASTAC (2000) is 1869 km^3 per annum but the amount of utilizable water is only approximately 1121 km^3. Water consumption for the year 2000 is estimated to be 520 km^3, of which irrigation takes 92 per cent. According to the prognosis, the annual need for water in 2025 will be 1027 km^3, i.e. roughly equal to the total amount of usable water and over 50 per cent of the total amount of renewable water resources.

Water distribution in the region is unequal (Figure 1.2). For example, in India, in the east there is plenty of water and in the west it is scarce. This situation has led to the idea of connecting the rivers running in the east-west direction via canals to form a network, which could get its water from Brahamaputra and Ganga. In this project we can see Bangladesh's water problems in the lower reaches of the Ganga. Currently, during the dry season, the river's run-off has already decreased by over 25 per cent. In addition to water shortage, salty seawater from the Bay of Bengal is penetrating deeper into the Ganga-Brahmaputra delta affecting all water use (Rabbi and Ahmad, 1997). In order to store water, India has built many reservoirs, among which the project on the Narmada River has raised many objections.

The source of life in Pakistan is the Indus. Most of the country's arable land is to be found in its valley. The total arable area is 22 million ha of which 74 per cent is irrigated. As in India, most of the potential agricultural land is already under cultivation. To feed the increasing population, more effective measures must be used in agriculture. Irrigated area has been increased by one-fourth since 1975 and the use of fertilizers has risen sevenfold. One of the biggest problems is that one-fifth of the irrigated area has become saline and one-sixth waterlogged. Extensive remedy projects are under way, yet the problems are huge.

The renewable water resources in Pakistan are 298 km^3 per year and water consumption was estimated to be 153 km^3 per year in 1990, of which irrigation accounts for 98 per cent (Frederiksen et al., 1993). The right of Pakistan to the water resources of the Indus has been secured by a treaty with India in 1960.

Bangladesh is situated in the delta area of the Ganga and Brahmaputra, in the Bengal lowland. The population density of the country is high: it

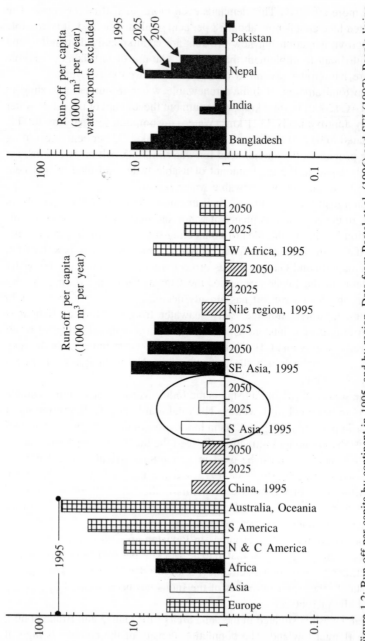

Figure 1.2: Run-off per capita by continent in 1995, and by region. Data from Postel et al. (1996) and SEI (1997). People living with less than 1000 m³ per capita per year tend to face a chronic water shortage.

has just passed 1000 persons per km^2. Agricultural land is 7.4 million ha. It is fertile, and climatic conditions allow even three harvests per year.

As population increases, the problem will be a shortage of arable land. The irrigated area has become threefold since 1970 and is now about 30 per cent. The use of fertilizers has increased in the same way as in India and Pakistan and has risen sevenfold.

The amount of renewable water resources in Bangladesh is, on average, 1387 km^3 per year and their use only 22.5 km^3. These figures, do not, however, give a real picture of the country's water problems. Floods in summer and drought in winter plague the inhabitants. Loggings in the Himalayas have aggravated the run-offs of the rivers flowing through the country. People have had to become accustomed to the fact that both floods and drought destroy crops. During the years 1973–87, average annual crop damages due to floods and droughts were 1.7 million tons and 1.5 million tons, respectively (Hasan and Mulamoottel, 1994).

In all these countries it is the quality of water that has become more problematic. Community wastewater is not properly treated. The same is true for industrial wastewater which contains heavy metals and poisonous compounds that constitute a serious hazard (Frederiksen et al., 1993).

A regional challenge is lack of cooperation between states. Without adapting the measures taken in Nepal and Tibet to the needs of countries in the lower reaches, water resources management will not achieve the best possible results (Upreti, 1993; Ahmad et al., 1994).

India, Pakistan, and Bangladesh have not been very peaceful for over half a century now, and Nepal, under pressure from its powerful neighbours, is not in an enviable position. International cooperation has been hard, and internal stability has not been forthcoming in any of the countries of the region; perhaps least of all in Pakistan.

Stockholm Environment Institute (SEI, 1997) classified Pakistan and India as high-stress countries with respect to water. Bangladesh and Nepal were classified as medium-stress countries.

South-east Asia and its Waters

South-east Asia consists of the countries of Indochina (Myanmar, Thailand, Lao PDR, Cambodia, Vietnam, Malaysia, and Singapore) and the Philippines and Indonesia. The region is 4.3 million km^2 (3.3 per cent of world total) and the population 500 million (8.3 per cent). Indonesia has 200 million, Vietnam 77 million, the Philippines 73 million, and Thailand 61 million. The population is predicted to grow to 600 million by the year 2025. Practically the whole area is pluvial, very fertile, and fecund. In spite of containing some important areas for global grain production and export such as Thailand's Central Plain, the region is a net grain importer.

The average GNP per capita was US$ 1300 per year in 1997 (2.3 per cent of world total). It is very unevenly divided. For instance, the GNP of Singapore's three million inhabitants is the equivalent of 14 per cent of the region's total GNP and is more than fifty-fold compared with the GNP of Lao PDR which has five million people. Differences within countries are also very noticeable (Drakakis-Smith, 1987). The GNP per capita for Jakarta, Bangkok, and Manila is one order of magnitude greater than the corresponding figure for the rural population. In some respects it explains why the metropolises of the region grow so rapidly.

Two important international rivers, the Mekong and the Salween, run through the region. Both have their source in Tibet. The Mekong flows through China (the Yunnan province) as a border river between Myanmar and Lao and then through Thailand and Lao. It continues to flow through Cambodia and runs from the Vietnam area into the South China Sea.

The river Salween runs through the region of China and Myanmar, and is a border river both between China and Myanmar and between Thailand and Myanmar. As far as the Mekong is concerned, since 1957 there has been the Mekong River Committee (from 1995, Mekong River Commission) supported by the United Nations and other international organizations. It has attempted to solve the regional water controversies with varying success. Its functioning has been made especially difficult by China's absence, and several national and international conflicts and wars in Vietnam, Cambodia, and Lao (Jacobs, 1995).

Among these countries Myanmar has experienced continuous internal disquiet and Indonesia has been volatile all the time. However, over the last few decades Indonesia has become more peaceful, but shows signs of underlying instability as the economy or the politics suffer from even modest crises. The biggest concentration of the population is the island of Java. Its area is 127,000 km² and the population 110 million. Its population density is almost 900 persons per km².

The region has 62 million ha of arable land (4.6 per cent of world total), of which 17 per cent is irrigated and grain production accounts for 49 million ha (7.1 per cent). Of the irrigated arable land Thailand has 37 per cent, Indonesia 23 per cent, Vietnam 14 per cent, and Myanmar 11 per cent. In Malaysia and Cambodia, on the other hand, agriculture is largely dependent on direct exploitation of rainfall.

Large and Dense Population

The expansion of human population is likely to continue at least for a couple of decades more. It occurs chiefly in the developing countries. The relative growth rates are gradually going down in most parts of the world, and this development is expected to continue. At present, the absolute growth is still linear, but is likely to start to decline relatively soon. Still, the population growth sets a heavy burden on the development possibilities of impacted countries, as well as on the planet's natural resources.

SOUTH AND SOUTH-EAST ASIA IN COMPARISON

South Asia is among the most densely habited areas of the world. India and Bangladesh have reduced their population growth rates dramatically but Nepal and in particular Pakistan have not. Pakistan's population is expected to double by 2025, which is a thought-provoking issue.

South-east Asia is on average far less populated than South Asia. However, there are certain areas which are extremely crowded, particularly the island of Java, most of the Philippines, and several river deltas, particularly those of the Red River, Mekong, Chao Phraya, and Irrawaddy.

The global share of the five regions mentioned in the Introduction is steadily growing; from 55 per cent in 1970 it is expected to reach 63 per cent by 2050. South Asia's share grows from 22 per cent to 23.5 per cent; population in the region hence grows 3.4–fold in eighty years. The corresponding figure for South-east Asia is threefold, the Nile basin 6.9, for West Africa 6.6, and for China twofold. Figure 1.3 shows the population projections for these regions.

IMPLICATIONS TO WATER

High population is a fundamental reason contributing to growing pressure on natural resources. Land is very limited and so are most other natural resources. Although many of the region's rivers are rich with water, the water availability per person remains relatively low.

The issue of population density does not receive enough attention in the debate on water. Massive agglomerations of people in parts of South and East Asia impose a unique stress on the environment. The associated problem of poverty and the structure of the agro-economic system imply that these regions are bound and will be bound to rely very much on local solutions in meeting their basic needs such as food supply. In parts of South Asia, the rural population density is higher than anywhere else.

Population growth per twenty-five years for various regions of the world

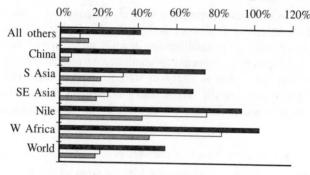

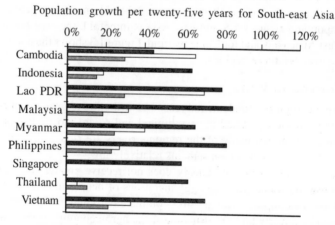

Figure 1.3: Population growth rates per twenty-five years. Population growth rates are going down, but doubles in one generation in many countries. The rates are adjusted for twenty-five years for the periods 1970–1995, 1995–2025, and 2025–50 (from the top). *Source*: UN (1994).

Urbanization is massive. The extremely land-scarce areas must be able to keep the mushrooming megacities alive.

Population growth, which is beyond the scope of water policies, is typically considered as an externality in the debate on water. It is true, however, that several issues which affect water development positively also help in population control. Many human development aspects belong to them—such as education, gender equity, poverty reduction, and so forth.

Massive Urbanization

In the global perspective, urbanization will perhaps be even a more problematic and momentous issue than population growth. Almost all population growth results in an exodus to cities. It is a big issue to most individuals in the coming decades, as well as when considered as a driving force in any aspect of humans and their environment, be it the nature, social development, or economics. Globally, rural and urban populations are now equal in size. By 2025, the rural population will not grow, yet the urban population will grow by 75 per cent.

SOUTH AND SOUTH-EAST ASIA IN COMPARISON

Nepal's urban population is expected to triple by 2025 (Figure 1.4). Bangladesh and Pakistan follow very closely. India, in contrast, will 'only' double the number of its urban population. India's urbanization development looks very similar to that of the emerging urbanizers of South-east Asia (Cambodia, Lao, Myanmar, Thailand, and Vietnam). Indonesia is expected to double its urban population, and the crowded towns and cities of the country will host over eighty million more people in 2025 than twenty-five years earlier.

In all these countries urbanization will constitute a huge challenge to any sector of the society, including the water sector. In 2002, South Asia had 400 million urban dwellers which might sound a small number in comparison to the 900 millions projected for 2025, although it is not a small number.

In South and South-east Asia, the rate of urbanization is comparable to the rate in China. In Africa, urbanization occurs with double the rate, challenging the societies in those regions much more than in Asia.

IMPLICATIONS TO WATER

The immense growth in the needs of water and food in cities will be rapid challenging all aspects of the water sector. Agricultural productivity must

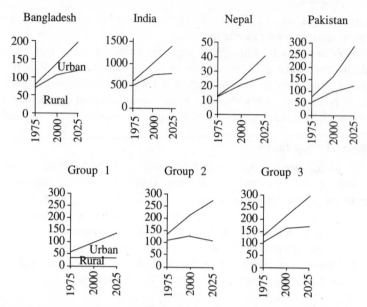

Figure 1.4: Rural and urban population in South Asia (above) and South-east Asia (below). The three groups for South-east Asia are: (1) Urbanized countries (Singapore, Malaysia, and the Phillippines); (2) Rapid urbanizer (Indonesia), and (3) Emerging urbanizers (Cambodia, Lao, Myanmar, Thailand and Vietnam). The urban population will continue to even quadruple in one generation. *Source*: UN (1994)

grow sharply. This cannot take place without massive improvement of irrigation efficiency. Arable land area as well as rural labour force will remain very much on the same level as before.

Growing urban centres will face enormous problems in ensuring safe water supply and appropriate sanitation to their inhabitants. In the recent past, South Asian countries have made great improvements in rural water supply (Figure 1.5) but not so much in urban water supply. Urban water infrastructure should be prioritized more than before. When it comes to progress in the sanitation situation, all the countries in the region have a long way to go before citizens can enjoy safe sanitation. The situation is worse in rural areas. Obviously, the water supply and sanitation situation should be developed hand in hand in order to ensure best results in terms of public health and environmental protection, but in South Asia, the development has not been in balance. A simple comparison to the situation in South-east Asia reveals this very clearly.

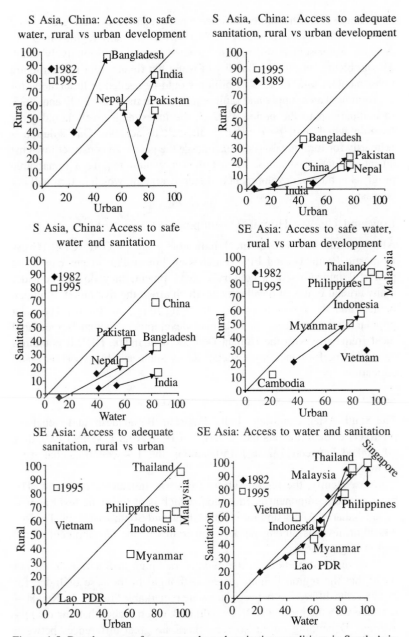

Figure 1.5: Development of water supply and sanitation conditions in South Asia in comparison to China and South-east Asia (*Source*: World Bank, 1999).

In South Asia, the domestic water supply requirements are growing up to 70–80 km^3 a year (SASTAC, 2000). Although many regions in South Asia are water-scarce, this required water volume is still relatively small. The problem is not the availability of water but finances for infrastructure investment. A total of US\$ 75 billion would be needed in India alone for appropriate water supply and sanitation between the years 2000 and 2025. In comparison to the period 1992–7, the investment level should grow fivefold which will be extremely difficult to achieve. One solution—primarily for rural areas—is to encourage the people to construct low-cost latrine sanitation systems. This, however, would require a remarkable progress in human development and empowerment, which in turn will not be easy.

Low and Uneven Human Development

Although the basic economical indicators such as GNP, GDP (Gross Domestic Product), and PPP (Purchasing Power Parity) per capita are powerful indicators of a country's development, they do not consider many crucial issues of development which affect the livelihood of human beings, and the possibilities for improving living conditions. The most popular alternative concept is human development, which is measured most frequently with the Human Development Index, HDI. It combines economic performance with social issues such as life expectancy and education.

SOUTH AND SOUTH-EAST ASIA IN COMPARISON

The South Asian countries—India, Pakistan, Bangladesh, and Nepal are relatively close to one another in terms of HDI. After UNDP (1999), they all ranked between 132 and 150, among the 174 countries that were included.

Consequently, the region suffers from low human development. The most critical component is education, which is at the same level as it is in sub-Saharan Africa and far behind China and South-east Asia. The low education level is among the key constraints of the development of the region (Figure 1.6).

The small differences in the country rankings in South Asia do not mean that the region is homogeneous, though. In the case of India, for instance, the differences between states are remarkable (Table 1.1). Kerala—the state with the highest HDI—would rank much higher than the country average. Depending on the source—there are great differences between the HDIs calculated by different authors—Kerala would be close to

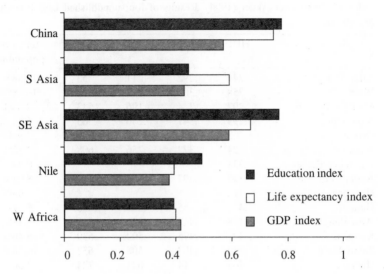

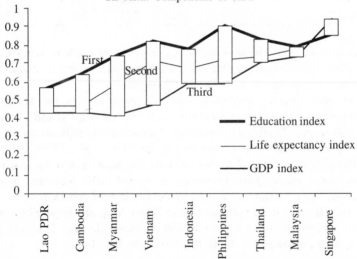

Figure 1.6: The three components of human development: education, longevity, and wealth. South Asia should be worried particularly about the low education level. *Source*: UNDP (1999)

Table 1.1: HDI in Indian states: a wide scatter. Disaggregated Human Development Index for India after Akder (1994). Results of four unpublished UNDP studies are shown

States	HDI[1]	HDI[2]	HDI[3]	HDI[4]	Share of population
Uttar Pradesh	.244	.292	.11	.53	16.88%
Bihar	.258	.306	.147	.503	10.48%
Madhya Pradesh	.297	.344	.196	.543	8.03%
Rajasthan	.299	.347	.246	.565	5.34%
Orissa	.3	.348	.224	.529	3.84%
Assam	.324	.372	.256	.608	2.72%
Jammu & Kashmir	.333				0.94%
Andhra Pradesh	.349	.397	.361	.589	8.07%
Himachal Pradesh	.413				0.63%
Gujarat	.417	.465	.566	.678	5.01%
West Bengal	.418	.467	.436	.641	8.26%
Karnataka	.427	.475	.502	.639	5.46%
Tamil Nadu	.436	.483	.508	.652	6.78%
Haryana	.467	.514	.624	.724	2.00%
Maharasthra	.484	.532	.655	.711	9.58%
Punjab	.538	.586	.744	.793	2.46%
Kerala	.603	.651	.775	.769	3.53%

Indonesia or even Thailand and Malaysia. The mid-northern states Uttar Pradesh, Bihar, Madhya Pradesh, and Rajasthan, to the contrary, fall among or even below the lowest HDI countries.

In each country of South-east Asia education is the strongest component of HDI, except Singapore where the economy is still stronger. In this respect South and South-east Asia are profoundly different.

Again, the enormous scatter among the HDIs evaluated by different authors gives a reason to be cautious in deriving too far-reaching conclusions on the basis of the disaggregated HDIs presented in this chapter. Same caution is justified to the whole HDI approach.

The HDI has been very positive in South Asia, and hopefully it will allow enhanced economic development in coming decades. However, the education component of the HDI is strikingly low (Figure 1.6).

IMPLICATIONS TO WATER

People-centred development provides many solutions, which cannot be met with the contemporary resource-based approaches to water resources development and management. Empowering the people to help themselves, raising public awareness and enhancing public participation are all important keys to overcome the hopeless financial situation in the region vis-à-vis the requirements. Active, aware, and empowered people are fundamental building blocks of IWRM in any country, also in South and South-east Asia.

South Asia has plenty of merits in the field of human-centred development. However, the severe deficiencies in human development hamper the power of these approaches seriously. Education would be the real booster to both economy and people-centred development. In South-east Asia, in contrast, people-centred development has a far better ground for growth.

Retarded Decentralization and Controversies in Empowerment

Along with the wave of globalization in trade, finance, and environmental issues, another worldwide force—namely decentralization—is reshaping development efforts everywhere. One of the basic ideas behind localization and decentralization is to enhance people's participation in politics and increase local autonomy in decision making.

SOUTH AND SOUTH-EAST ASIA IN COMPARISON

South Asia has a strong tradition of paying attention to the self-rule of the poor. The most important approaches to empowerment have strong links to the region—from Gandhi to Sen and Grameen Bank and beyond. Rural development has been seen to be very important in all the region's countries.

However, in many ways the countries are very centralized—in financial terms far more than China for instance. Bangladesh is particularly centralized—the local authorities govern only 3 per cent of national revenues whereas in China the share is 50 per cent. Centralization is prone to hinder regionally balanced development in South Asia, and may lead to centralized solutions to the cost of localized ones. South-east Asian countries are also very centralized. Only in Malaysia the provinces get 15 per cent of national revenues. In all other countries their share remains between 3 and 8 per cent.

The South Asian situation is changing. A few decades ago many ideals were along the lines of self-esteem, reality was a complex mix of power structures, and empowerment was not well developed. Recently, economic and social liberation, which has gone to some extent hand in hand with raising religious-traditional influence, has offered increasing possibilities particularly to the middle class. The history of many South-east Asian countries has been very militant, but the situation seems to be changing as well, towards the same direction as in South Asia.

The gender issues and empowerment of women are a big challenge to both regions, but to South Asia in particular. Gender empowerment indicators for South Asia are very low, lower than any other part of the world.

IMPLICATIONS TO WATER

The South Asia Water Vision states: 'Role of government will change with appropriate degrees of centralization and decentralization of water resources planning, development, and management. Privatization can lead to efficient delivery of services to consumers at least cost; to financially self-sustaining organizations; to have water treated as an economic good; to correlation of the levels of service to the prices charged; and, to effective cost recovery.'

This tendency is welcomed and progress in decentralization is necessary. When it comes to privatization, problems might arise owing to the fairly low level of cost return that prevails in the water sector investments in the region. This issue is commented in more detail under the title 'Economic Underdevelopment'.

Empowerment, however, should be much more emphasized in the water sector than is done at the moment. Civil society organizations are functioning better and better in the region, and in fact, the civil society here might be working better than in most developing regions of the world. However, large masses are still not in a position to be able to influence living conditions in favour of themselves. Disparities in gender, education, economy, and consequently empowerment and many other respects are enormous.

The Water Vision (SASTAC, 2000) pays justified attention to the issue of gender and to the need for empowering women. Gender sensitive water management is emphasized as well. Rahaman et al. (2002) propose the following six initiatives to be urgently taken in order to enhance gender equity in water issues:

   (i) Ensure women ownership to land;
  (ii) Need for a powerful regional umbrella organization for women;
 (iii) Facilitate the voice of women in all planning, management, and decision-making processes;
  (iv) Ensure education for female, especially in rural areas;
   (v) Recognition of women as income producing agents; and
  (vi) Facilitate access to all natural resources for women.

Slow Globalization and Regional Integration

No other concept is as hated or loved at the same time these days as globalization. Broadly speaking, it means opening of the gates and removing boundaries between nations. While the basic idea is grand and the underlying tendency is inevitable in the contemporary world, plenty of contradictions and side effects are obvious.

Arguments such as comparative benefits due to international division of labour and substitution of commodities are often used to back the benefits of international trade. Tariff barriers have gone down almost everywhere, allowing enhanced conditions to trade across borders. Yet, the protection of poor economies is highly justified, knowing the extreme disparities in the world economy. Besides, the wealthiest economies such as US and the European Union (EU) are all but tolerant in allowing foreign products to enter their markets.

World trade doubled in only ten years between 1987 and 1997 (World Bank, 1999). The ratio of trade to world's total purchasing power parity adjusted GNP grew from 20.6 to 29.6 per cent. Perhaps a still more sensitive indicator of the integration of an economy to the world market is the share of gross private capital flows across the borders of the economy to purchasing power parity. This indicator value grew from 7.0 per cent in 1987 to 12.7 per cent in 1997, if calculated over the world.

Whatever the attitude for or against globalization, most people agree that at least regional cooperation would be beneficial.

SOUTH AND SOUTH-EAST ASIA IN COMPARISON

South and South-east Asia have witnessed an opening up of economics in the past decade. However, South Asia is still far behind all other of the world's macro-regions in terms of integration into the world market. In 1997, the five study regions accounted for 2.76 per cent of the world's private capital flows (World Bank, 1999). South-east Asia accounts for 2 per cent (among which Singapore for 1.6 per cent, China for 0.45 per cent, and South Asia for 0.13 per cent). The Nile region's share is 0.08 per cent, whereas West Africa's share is 0.06 per cent.

Tariff barriers have gone down rapidly, but capital flows are still very low. However, the capital flows are already in modest growth. Foreign trade plays an exceptionally low role in the region's economies (Figure 1.7). Consequently, the region can expect less external investments and foreign capital than most other parts of the world. It must largely rely on its own resources. The problematic relations of India and Pakistan cast a dark shadow on the integration process of this region.

Foreign trade (0% of PPP GNP)

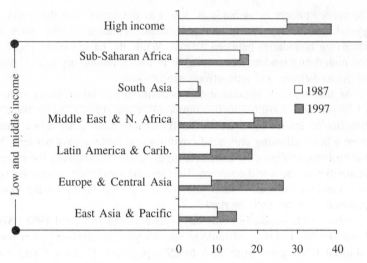

Private capital flows (% of PPP GNP)

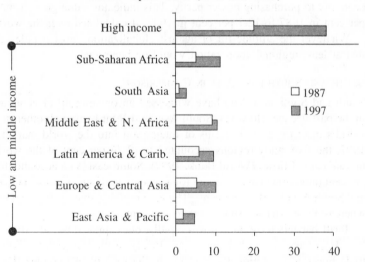

Figure 1.7: World's regions: Integration with the world economy. Trade in goods (imports plus exports) and gross private capital flows as a share of purchasing power parity adjusted GNP. *Source*: World Bank (1999)

IMPLICATIONS TO WATER

The countries of South Asia are expected to still rely on self-sufficiency in many basic commodities for a long time. This has important implications to water management since over 90 per cent of the region's used water goes to agriculture. Along with urbanization and population growth, the water used in agriculture is expected to grow unless attempts at saving water become far more efficient than at present.

The role of foreign investments being very low in South Asia, ultimately the countries must be able to find the enormous sums needed from their own economies. The foreign capital flows are minor at the moment, and are growing only very slowly. After SASTAC (2000), the average project-level investments required on water for the region's countries are as given in Table 1.2. It must be noted that Nepal's figure is clearly unrealistic; it is more that one-third of the country's GNP.

Table 1.2: Investment needs to water projects: annual averages by 2025 (SASTAC, 2000) and their share of the countries' GNP (of 1999 after World Bank, 2001)

	Annual investment need	
	Billion US$	As % of GNP
India	5.00	1.13
Nepal	1.80	34.60
Pakistan	1.04	1.60

The private sector is often called for in this context, but it must be remembered that cost recovery in the water sector is very low in South Asia. In irrigation and multipurpose projects it is 13 per cent while in small-scale irrigation schemes it is only 2 per cent. Hardly an attractive target for private investment.

The tense regional atmosphere is not conducive to regional integration. Many waters are shared and should be managed internationally. Presently, despite political tensions, many water-related institutions such as the Indus Water Treaty are functional and contribute to confidence building. Bangladesh as a downstream country would wish to have more to say on international water issues, particularly with regards to the Ganga-Brahmaputra-Meghna river system. The operation of the Farakka Barrage has remained a problematic issue. The need to stimulate the regional cooperation is highlighted in the Water Vision (SASTAC, 2000).

South-east Asia has witnessed remarkable regional integration in the past, particularly in terms of ASEAN-led economic cooperation. Trade and investments are much higher than in South Asia.

Malfunctions in Governance and Policy Environment

In a large number of the world's countries, the governance system suffers from serious malfunctions including corruption. The legislative systems are handicapped by overlaps and inconsistencies, and many laws are not implemented. National water policies may exist but are not very effective. Decentralization and increasing participation of the private sector are considered by some as possible solutions to government malfunctioning. The results thus far from the region are very mixed to draw any definitive conclusion.

SOUTH AND SOUTH-EAST ASIA IN COMPARISON

International comparisons do not give a very positive image of the governance situation of the South and South-east Asian states. Bangladesh remains at the last position in the international comparison of corruption by Transparency International (see e.g. Transparency International, 2002; Figure 1.8). In this light a lot needs to be done to improve the functioning of the governance system in the subcontinent.

IMPLICATIONS TO WATER

As in many other parts of the world, planning in the water sector is changing at present from top-down technocratic approach to bottom-up grassroots approach (Ahmad et al., 2001). The approaches are more participatory than before, and partnerships between public and private operators are called for.

The Water Vision (SASTAC, 2000) pays plenty of attention to various issues in institutional development, governance, and policy environment. This is perhaps one of the boldest sections of the vision. The following quotation characterizes the spirit in which the Vision discusses the governance issue: 'The issues of governance, policy and institutions are so critical that without addressing them in adequate measure in each of the countries, integrated development and management of the water resources for realising a sustainable water vision in 2025 will only be an academic exercise. Many reforms are needed in governance and institutions.'

Policies as well as laws may exist but very often they never become effective. Legislation suffers from overlaps and needs a profound reform. Governments are too heavily involved in controlling the resources even at the micro level. Decentralization is seriously lagging behind, as is the

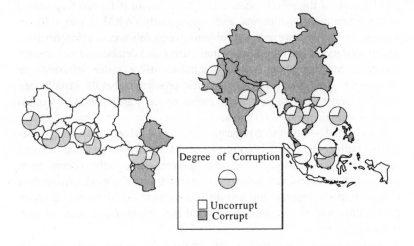

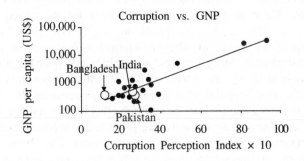

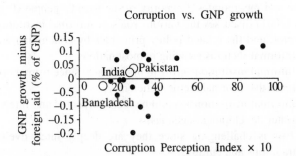

Figure 1.8: The 2002 Corruption Perception Index (CPI). Degree of corruption grows towards south, and correlates well with GNP per capita.

Source: Transparency International (2002)

involvement of the private sector. The water sector is far too fragmented in the government institutions, and consequently IWRM is very difficult to implement. There are serious problems in people's access to information, which hinders the development of transparent and democratic governance practices. Substantial amount of capital-intensive water infrastructure investments have been made, but a typical problem is that the installations are deteriorating due to inadequate maintenance. Participation of women in water management is very low.

As a partial solution to governance problems, privatization and public-private partnerships have been proposed. This policy is justified in many cases. Yet, in situations where the governance system suffers from serious malfunctions, this policy has serious traps. If a weak government cannot properly regulate the private sector, there could be public sector partnerships with the informal sector and the private sector, as is the case in many countries.

A crucial issue that is missing largely from the contemporary debate on water governance is the informal sector and consequently the informal institutions. These institutions provide the rules for the society. Their various functions range from providing legislative, juridical, and administrative framework to different informal aspects such as culture, religion, and ethnicity. Along with rapid urbanization, economic liberalization, and other transitions, the various roles of informal institutions are increasingly being emphasized in development programmes, but not yet properly in water agendas or visions. Policies that are based on promotion of public awareness, grassroot activities, participatory approach, and so on are often targeted at least partly to the informal sector. This sector, leaning largely on informal institutions, grows rapidly in developing and transitional countries, and incorporates a majority of the world's people (Figure 1.9).

Varis (2001) discusses the various roles of informal institutions in the water sector, and the related policy principles by concluding as follows:

(i) Informal sector is growing in most nations;
(ii) Informal institutions grow in importance, since formal ones do not reach the informal sector sufficiently;
(iii) Informal institutions should be more respected and integrated into water development agendas;
(iv) This is challenging since they are deeply interwoven into the traditions and culture;
(v) Their positive aspects should be supported, and their negative sides such as corruption, bribery, local mafias, etc., should be set under control.

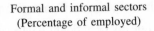

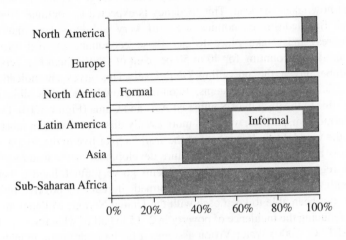

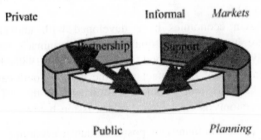

Figure 1.9: Role of the informal sector. *Above*: The informal sector employs two-thirds of the Asian labour force (Charmes, 1998). *Below*: Positive interconnections between public, private, and informal sectors. The public and private sectors should be able to work in partnership, and the informal sector should support the public sector (Varis, 2001)

Economic Underdevelopment and Widespread Poverty

Poverty reduction has found its way to all development agendas. The definitions and the number of those living in poverty vary greatly, but roughly one-fifth of mankind is usually classified as being poor. One of the eight Millennium Development Goals of the United Nations is to halve the incidence of poverty from the 1990 level by the year 2015.

South and South-east Asia in Comparison

Pakistan has traditionally been South Asia's strongest economy, but India has now taken the lead. This tendency is expected to continue. Nepalese and Bangladeshi economies are still very aid-dependent and much progress must take place to get those economies on feet. Services dominate, accounting for 40 to 50 per cent of GNP, which is a very high number given the prevailing GNP level. This gives an indication of inefficient economic systems. Economic growth, though visible is far inferior to the growth of South-east Asia and China (Figure 1.10). Despite enormous disparities, wealth is more evenly distributed than in most parts of the world. Still, 36 per cent of the world's poor live in the region which poses a serious problem to the future development of the water sector in the region—poverty and water problems go very much hand in hand in South Asia. Though strides have been made in reducing poverty, progress has not been able to keep pace with the United Nations Millennium Goal of reducing the incidence of poverty in 2015 to half of what it was in 1990. SASTAC (2000) Water Vision assumes a far more dramatic reduction in poverty (Figure 1.11).

South-east Asian economies have developed with uneven rates. Singapore's share has more than doubled, and Myanmar's share has shrunk drastically. Indonesia had good times till the early 1980s, but after that, development has not kept intact with the rest of South-east Asia. The Philippines has also some trouble in following the others. The split of the regional economy is clear from the discussion below:

The losers: The poor, communist or post-communist economies of Lao PDR, Cambodia, Vietnam, and Myanmar, with 140 million people and GNP per capita around US$ 300. Their proportion of regional population is 28 per cent. Their share of regional GNP has dropped from 20 per cent to 4 per cent between 1970 and 1997, although some positive signs can be seen in the context of Vietnam.

The in-betweens: The lower-middle income isles, Indonesia and the Philippines, have 275 million people with a GNP per capita approaching US$ 1200. Their share of the regional GNP has grown slightly, from 43 per cent to 45 per cent. They have 55 per cent of the regional population.

The winners: The emerging economies Singapore, Malaysia, and Thailand, with GNPs per capita US$ 33,000, 4500, and 2700, respectively. Their share of regional economy has grown from 37 to 51 per cent. Population in these three countries accounts for 17 per cent of the regional total.

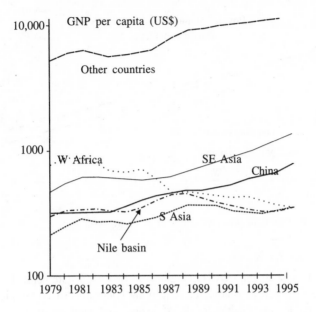

People in absolute poverty % of World Total (1.44 billion)

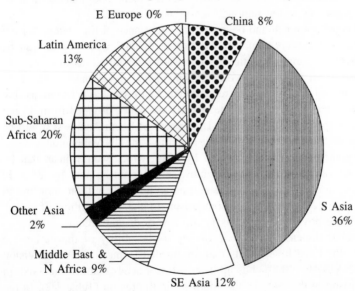

Figure 1.10: Economic growth in South Asia vis-a-vis other regions. South Asia's economic growth is slow and poverty problem is massive (World Bank, 1997).

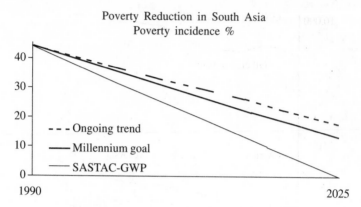

Figure 1.11: Trends in poverty reduction in South Asia.

IMPLICATIONS TO WATER

The South Asia Vision (SASTAC, 2000) starts with a statement: 'Poverty in South Asia will be eradicated and living conditions of all people will be uplifted to sustainable levels of health and well-being, inter alia, through coordinated and integrated development and management of water resources in the region.'

Whereas optimism and belief on the importance of the own sector are important, it is very doubtful whether a regional water vision can be considered as being realistic if it is based on assumptions that are very different from most of the other distinguished development prospects. The historical trend in poverty reduction in South Asia has been remarkable from the year 1990 to 1999; the poverty incidence dropped from 44 per cent to 37 per cent. At this rate, 18 per cent of South Asians still will be poor in 2025. With a more optimistic scenario—the UN Millennium Goal— poverty should be reduced to half by 2015, which would mean that by 2025, less than 14 per cent of South Asians would be poor by 2025. In absolute numbers, in 2015 there will be around 340 million poor and in 2025 about 270 million, in comparison to the 500 million of today. Eradicating poverty completely is still a very distant dream.

Another dimension of the economy is the ability of the nations to finance the huge investments that are needed to meet the Water Vision goals. SASTAC (2000) ends up in investment needs that are excessive in comparison to the GDP of the countries in the region (Table 1.2). In the case of Nepal, one-third of the national GNP would be needed for investments in the water sector. There is a discrepancy of one order of magnitude to any realistic investment level. The low economic growth rate in comparison to many other developing regions does not help the situation, nor does the low level of integration to the global economic system.

In the case of South-east Asia, the economic winners and perhaps also those in between can be assumed to be able to meet the millennium goals of poverty reduction. However, the other countries will definitely have it very difficult in this regard.

Food Insecurity

Approximately every sixth human being suffers from food insecurity. Global food security projections suggest better days to come. However, the optimism is largely based on the assumption that low-income countries will increase their food production and imports. This means, that their economies should grow steadily, and food markets should be stable.

SOUTH AND SOUTH-EAST ASIA IN COMPARISON

The regions have improved its food production systems remarkably in the past decades. However, malnutrition is still widespread, and part of the progress is eaten away by the rapid population growth. Rapid urbanization, climatic effects such as droughts, and many social disparities continue to cause food security problems to one-fifth of the region's population, which is an alarming number. In early 1990s, more than 800 million people were undernourished in the world, among which 180 million were children (Figure 1.12). In South Asia, more than 50 per cent of children are undernourished. In South-east Asia, the situation is similar except in the economic winner countries (see the previous section).

IMPLICATIONS TO WATER

Self-sufficiency is important in the region, both locally and nationally. Irrigation is very often needed to increase productivity to facilitate economically rewarding farming. Impact of urbanization is large: market-driven agriculture must grow rapidly. Lack of purchasing power is a serious obstacle.

Agriculture's share of water withdrawals in South Asian countries is very high; it is 86 per cent for Bangladesh, and for the other countries it is well above 90 per cent. The water use efficiency, however, is fairly low. The problem with Indian and with almost all South Asian agriculture has been its ineffectiveness. According to Pike (1995), yields from India's

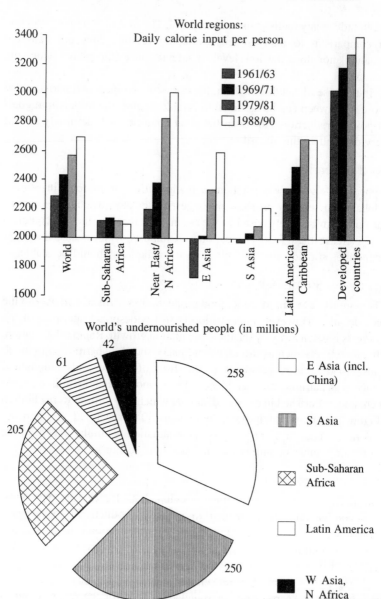

Figure 1.12: Food situation in the early 1990s after IFPRI (1997). *Above*: Daily calorie input per person in different regions of the world (Alexandratos, 1995). *Below*: World's undernourished people

irrigated areas are only about 50 per cent of the respective yields per hectare in other parts of Asia. The great challenge of the region is to improve the efficiency, as far as soil and water use are concerned, by emphasizing training, technology, and irrigation systems management (ICID, 1995). Pakistan leads the world globally in used water per produced unit of GNP (Figure 1.13). The country's economic structure is based on enormous water consumption, which produces very little wealth. Thailand is an example of the reverse. It is a strong agricultural country with massive irrigation infrastructure. Despite heavy investment in agriculture, it has at the same time diversified its economy and produces now eight times more wealth per capita from its economy that uses less than half the amount of water per person, in comparison to Pakistan.

With the amount of water used for agriculture, there should be better food security. Food security is not only an agricultural issue. It is a social question closely related to poverty.

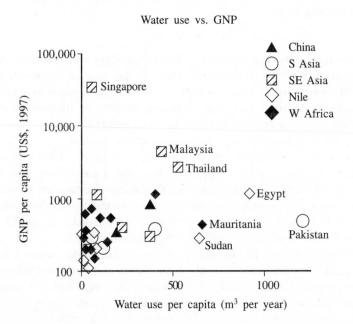

Figure 1.13: Gross National Product (GNP) vs. water consumption
Source: World Bank (1999)

Variations and Changes in Climate

The globe's climate is subjected to variations due to innumerable reasons. The hot topics at present in research pertaining to climatic change are whether human activities have changed global climate, how this has happened and will continue to evolve, what the most important consequences will be, and how future changes should be mitigated with present actions. These changes are most frequently attributed to increasing concentrations of greenhouse gases and aerosols in the atmosphere, as well as to extensive changes in the usage of land leading to deforestation and desertification.

Such changes, however, are anticipated to vary much with respect to local and regional weather conditions. Some evidence suggests that the changes in variations are much more significant than those in the mean behaviour of the climate. In fact, extremes such as floods, droughts, storms and their frequency are often critical to water resources management, and changes in their pattern or frequency can be very influential to the economy and society.

SOUTH AND SOUTH-EAST ASIA IN COMPARISON

All the five study regions listed in the Introduction are located in the equatorial belt. This belt covers the range of the seasonal movement of the intertropical convergence zone. In this zone, the trade winds from both hemispheres confront each other, causing rainfall. Its location varies seasonally, being in extreme north during the northern hemisphere summer, and in extreme south in the southern hemisphere summer.

This belt is subjected to highest uncertainties in the global climate change projections (IPCC, 2001), and in a few decades from now, it will be affected more by other global changes—population growth, urbanization, industrialization, and political-economic transitions—than other parts of the world. This makes the economies particularly sensitive.

South and South-east Asia are fully dependent on the monsoons. Particularly sensitive to climatic variations is the border zone of the monsoons. This zone crosses Pakistan and North-east India, and follows the Himalayas from the south all the way to Assam. South-east Asian countries, however, are not particularly sensitive in this respect.

IMPLICATIONS TO WATER

The climatic component does not receive much scrutiny in the present water debate of South Asia. SASTAC (2000) mentions it in a general way as a driving force: 'It is also expected that the impact of climate change and sea level rise will have significant implications concerning both the temporal and spatial availability of water. General circulation modelling revealed that monsoon precipitation will increase by 11 and 27 per cent under moderate and severe climate changes respectively, while there will be no appreciable winter precipitation.'

However, the vision obviously considers such intuitively precise figures as unrealistically confident, since later it continues: 'Climate change and sea level rise and their consequences are not precisely predictable but can be serious in parts of the region requiring adaptation to the new situations.'

Now, no real implications are discussed, except some reference in the case of flood and drought management.

Here again, a more cautious attitude, taking into account the severe threats the climate poses to the region, would be justified. Bangladesh is one of the most vulnerable countries of the world to climate extremes. However, also the other countries of South Asia are very sensitive—be the problems associated with severe droughts, floods caused by monsoon storms or snowmelt, or other issues such as sea level rise. Also a thorough understanding to the region's own potential impact to the climate—caused by rapidly growing emissions and massive changes in land use—as well as their relationship to flood and drought management are called for.

The present status of climate projections does not unfortunately allow a concrete basis to build up other policies besides making societies less sensitive and vulnerable to any foreseeable event. It is very doubtful to base any real-world policy choices in present projections pertaining to climate change. The very basic needs of the billions that are potentially affected by changes—water, food, housing, social fabric—are in question. The models give an overly optimistic view of the adequacy of the knowledge, despite their consistent development. In practice, precautionary policies are necessary.

Conclusions

This short overview of the critical externalities to water resources development in South and South-east Asia can only cast a very brief glance over this huge issue. However, the water sector should be better aware and conscious of the present state of and future expectations with regards to these externalities. The water sector could even on its own make earnest attempts to foresee and reduce their gravity by rightly targeted policies. What these policies could be is too broad an issue to be scrutinized in any single study. However, seeing the water issues in the broad framework of other development issues such as the ones discussed in this chapter—and integrating the visions and policies of the sector—would be the way to go towards the future in South and South-east Asia.

The questions raised in the beginning were whether the Water Vision is realistic in terms of these externalities. The comments raised on the basis of the analysis are summarized in Table 1.3.

It could be argued the Vision should be optimistic. However, it is doubtful whether the Vision is realistic, since it is not based on enough thorough analysis. Considering the diversity, extent and magnitudes of the problems faced by the countries of the region, fulfilling the millennium development goals is going to be a very difficult task.

The Vision's approach should be more human-centred. The people, the civil society, and the governance with the informal and formal institutions constitute the machinery which will, to a significant extent, determine the quality of the final results. It is important, of course, to set sector-specific targets and formulate action plans for the purpose of reaching those goals. But it is necessary to relate these actions to the needs of the society and try to change the rules and educate its stakeholders at the same time. Otherwise not much can be achieved.

Table 1.3: Comments on the South Asia water vision: Towards a more human-centred water vision

Issue	Water vision	Suggested enlargement of the vision
	'Water resource centred'	'More people-centred'
Population	Passive attitude—take forecasts and speculate how to meet needs	Try also to reduce population growth through paying more attention to the people
Urbanization	Passive attitude—take forecasts and speculate how to meet needs	Try to change the modus of urbanization from rural push to urban pull
Low and uneven human development	The situation in large is accepted passively and some very specific actions are proposed such as training courses for farmers	The underlying root course for many water problems lies in low human development. This should be understood.
Retarded decentralization and controversies in empowerment	Emphasizes the issue but does not go into the profound political and social roots of the problematique	Empowerment and decentralization issues are much more profound in character and major attitude shifts are necessary
Slow globalization and regional integration	There is a concern on international food and water markets for instance. Now, investments are badly needed, and foreign input could be important. On the other hand regional cooperation particularly in terms of regional sector institutions and partnerships is encouraged.	The region has the largest Anglophone population in the world. Culturally, the region is exceptionally strong. There should be possibilities to activate and empower large populations to allow in-depth integration of the subcontinent to the global community.
Malfunctions in governance and policy environment	Governance system is expected to 'commit itself for achieving the vision'. Government's role is emphasized as 'regulator and facilitator' of water services. Profound reforms are seen as a prerequisite for achieving the vision.	Governance systems of the region are among the most corrupt in the world. It is agreed that profound changes in governance are necessary, not just a 'commitment to the water vision'.
Economic underdevelopment and widespread poverty	The vision statement commences with the phrase 'Poverty in South Asia will be eradicated...' but does not propose any	The population with food security problems, with chronic public health problems, with low education and the poor are not completely

(Contd.)

Table 1.3 (Contd.)

Issue	Water vision	Suggested enlargement of the vision
	other but water-sector related actions. This will not be enough	but to a large extent the same group of marginalized people, particularly in urban areas. It is naive to assume that poverty be eradicated and then we provide water supply and sanitation to all, and reduce the morbidity to water-borne diseases to zero
Food insecurity	The link between food insecurity and poverty is recognized. Economic growth and technological advance are suggested, as well as additional water storage volumes to overcome food insecurity problems	The human dimension is again forgotten. Education, public awareness, gender equity, and, in general, the ability of people to tackle the nutritional issues from bottom up are very important
Variations and changes in climate	Climate variability and possible changes in climate are not considered as a driving force	The climate-related risks, including possibly worsening floods or droughts or both, should be on the agenda, given the very sensitive monsoon climate of the region
Environmental degradation	Environment is considered in a many-sided way in the vision, but the social and economic links are missing	Many environmental problems are linked closely to social problems and instabilities. The informal sector, rural poor, poverty-trapped urban communities such as slums etc. are an important source of non-governed environmental problems Moreover, there are obvious needs to restructure the macroeconomy of the region by taking the environmental considerations into account. One example is the enormous water use in many parts of the region in comparison to produced GNP

References

Ahmad, Q.K., A.K. Biswas, R. Rangachari, M.M. Sainju, 'A Framework', in Q.K. Ahmad, A.K. Biswas, R. Rangachari, M.M. Sainju (eds). *Ganges-Brahmaputra-Meghna Region: A Framework for Sustainable Development*: 1–29, Dhaka: The University Press Limited, 2001.

Ahmad, Q.K., B.G. Verghese, R.R. Iyer, B.B. Pradhan, S.K. Malla (eds), *Converting Water Into Wealth: Regional Cooperation in Harnessing the Eastern Himalayan Rivers*. Kathmandu: Institute for Integrated Development Studies, 1994.

Akder, A.H. 'A means of closing gaps: Disaggregated Human Development Index'. Occasional Paper 18, New York: United Nations Development Programme 1994.

Alexandratos, N., World Agriculture: Towards 2010—An FAO Study, Rome: FAO, and Chichester: Wiley, 1995.

Charmes, J., 'Informal sector, poverty and gender: A review of empirical evidence', Washington DC: Background Paper for the World Development Report 2001. World Bank, 1998.

'Comprehensive Assessment of the Freshwater Resources of the World', Stockholm: Stockholm Environment Institute, 1997.

Drakakis-Smith, D., *The Third World City* London: Routledge, 1987.

Frederiksen, H.D., J. Berkoff, W. Barber, '*Water Resources Management in Asia*', Vol 1, Main Report, World Bank Technical Paper 212, Asia Technical Department Series Washington D.C.: World Bank, 1993.

Gleick, P.H. (ed.), Water in Crisis, New York: Cambridge University Press, 1993.

Hasan, S., G. Mulamoottel, 'Natural resources management in Bangladesh', *Ambio*, 23, 1994, pp. 141–145.

'Human Development Report 1999', United Nations Development Program/ Oxford University Press, 1999.

IPCC, *Climate Change 2001: The Scientific Basis Cambridge*, Cambridge: Cambridge University Press, 2001.

'Integrated Water Resources Management', Technical Advisory Committee Background Paper No 4, Stockholm: GWP, 2000.

'International Program for Technology Research in Irrigation and Drainage: An ICID Vision', New Delhi: International Commission on Irrigation and Drainage, 1995.

Jacobs, J.W. 'Mekong Committee: history and lessons for river basin development', *Geogr. J.*, 161, 1995, pp. 135–48.

Pike, J.G., 'Some aspects of irrigation system management in India', *Agricultural Water Management*, 27, 1995, pp. 95–104.

Postel, S., G.C. Daily, P.R. Ehrlich, 'Human appropriation of renewable freshwater', *Science*, 271, 1996, pp. 785–88.

Rabbi, M.F., E. Ahmad, 'Environmental degradation of the southwest region of Bangladesh and need for a barrage on the Ganges', Proceedings, International

Conference on Large-Scale Water Management in Developing Countries: EI42–EI49, 20–23 October 1997, Kathmandu, Nepal.

Rahaman, M.M., O. Varis, T. Kajander, 'On the search for holistic water management and sustainable environment through women participation in South Asia', Second South Asia Water Forum, 14–16 December 2002, Islamabad.

'South Asia—Water for the 21st Century: Vision to Action', Global Water Partnership/South Asia Technical Advisory Committee, 2000, Aurangabad.

'The World Food Situation, Recent Developments, Emerging Issues, and Long Term Prospects', Washington D.C.: The International Food Policy Research Institute, 1997.

Upreti, B.C., *Politics of Himalayan River Waters—an Analysis of the River Water Issues of Nepal, India and Bangladesh*, Jaipur: Nirala, 1993.

Vakkilainen, P., O. Varis, 'Will water be enough, will food be enough?', *Technical Documents in Hydrology*, 24, Paris: UNESCO/IHP, 1999.

Varis, O., 'Informal Water Institutions', Proceedings of the IWA 2nd World Water Congress, 15–19 October 2001, Berlin.

Varis, O., 'The Nile Basin in a global perspective: natural, human, and socioeconomic resource nexus', *Water International*, 25, 2000, pp. 624–637.

Varis, O., P. Vakkilainen, 'China's 8 challenges to water resources management in the first quarter of the 21st Century', *Geomorphology*, 41(2–3), 2001, pp. 93–104.

World Development Indicators, Washington D.C.: The World Bank, 1997, 1999, 2001.

'World Population 1994', New York: United Nations, 1998.

2

INTEGRATION IN BITS AND PARTS: A CASE STUDY FOR INDIA[1]

A.D. Mohile

Introduction

Integrated Water Resources Management (IWRM) is by now well recognized as a set of principles appropriate to the current water problems. This chapter reviews the status of and experience with IWRM as applied to water management in India, in particular the various facets of integration intended in IWRM. This is done both according to the facets of integration and according to basins having a rich experience.

It is found that although the principles are well recognized, and applied, only a partial integration has resulted. A well-managed basin, which could be considered as a showcase for IWRM, does not emerge from the analysis, and this appears to be a somewhat distant target.

Even while accepting the process of IWRM as the guiding principle for future water management, the chapter brings out the need for a dynamic interpretation of the principles of IWRM, to allow for integration going beyond the concept of IWRM, so as to include inter-basin integration as is perhaps desirable in India, through long-distance water transfers.

This analysis also brings out the need for considering the spatial limits of integration. It brings out that for some international and interstate basins, solutions, which do not fit into the principles of IWRM, have been found and have worked rather satisfactorily. It suggests that both the sovereignty of nations as also the autonomy of states in a semi-federal

[1] A preliminary draft of this paper was discussed in the Bangkok consultation about IWRM in South and South-east Asia (Dec. 2002). The comments resulting from there, and in particular, the suggestions of Asit K. Biswas and Olli Varis are gratefully acknowledged.

set-up may require special mechanisms and considerations in resolving conflicts, and an analysis of such situations may provide an understanding about the limits of IWRM.

Purpose and Scope

THE BACKGROUND

IWRM, which is defined by the GWP (2000) as a 'process which promotes the coordinated development and management of water, land and related resources, in order to maximise the resultant economic and social welfare in an equitable manner without comprising the sustainability of vital ecosystems' is being considered as the overriding set of principles appropriate to current water management problems, characterized by water stress through growing population, increasing pollution, and a crisis in water-related governance. (These 'Dublin Principles' of 1992 themselves perhaps need modifications based on the Johannesburg declaration of 2002 but in the absence, the IWRP concepts have to continue to be classified in accordance with these.)

It is important that IWRM is looked at as a process under evolution and not as a one-time goal to be achieved. Even while the set of four principles would normally have a universal acceptance, the environments under which the principles are to be put in practice may have severe constraints; and although the management policies could to some extent be tuned, through a process aimed at removing such constraints, this may be very difficult.

THE SCOPE

In this context, this chapter reviews the status of IWRM in India with reference to available cases. It is seen that the principles of IWRM have been considered in the past, in planning development and management of water resources, but could be practised only in parts. It analyses the causes, discusses the shortcomings of the Indian situation, both for the present and projected future changes, and gives some suggestions about possible policy changes.

Introduction to the Indian Water Sector

India lies roughly between 7° N to 36° N and 10° E to 93° E. It has an area of about 3.29 million km². The topography of India consists of the high and lofty Himalayan mountains in the north, the plain valleys of the Indus,

Ganga, and the Brahmaputra south of the Himalaya, and a large peninsular plateau and hills, and narrow coastal plains.

The precipitation in the country depends mostly on the south-west monsoon, but the north-west monsoon also causes rains in the southern part of the eastern peninsula. The average annual precipitation is about 1215 mm but the spatial variation of normal is very large, from less than 100 mm in the western deserts, to 11,000 mm in parts of the north-east.

The major basins/basin group of India are shown in Figure 2.1.

The average annual natural freshwater resources of India has been estimated at about 1950×10^9 m^3 (i.e. 1950 km^3). Of this, about 690 km^3/ yr is considered as utilizable from surface sources and another 432 km^3/ yr as utilizable from the groundwaters.

In general, the Ganga-Brahmaputra-Meghna (GBM) basin (and in particular the Brahmaputra and Meghna arms) have much larger water resources. Other basins of the northern peninsula, flowing to the east, as also the numerous small west-flowing rivers of the peninsula are comparatively better off in water resources. The skewed nature of the water resources is brought out in Table 2.1.

A considerable water development has already taken place. The present annual withdrawal for various purposes would total to around 550×10^9m^3, and the present consumptive use would be around 300×10^9m^3. The irrigation sector is the predominant consumer of water. As per the report of the National Commission (1999), the annual withdrawal requirements, by 2050, could be between 973×10^9m^3 and 1180×10^9m^3, and about 40 per cent of these would come from the groundwaters. These withdrawal levels could become possible after resorting to reuse of returned waters, considerable management improvements, and continued storage-based development.

The Indian Water Policies and the IWRM Principles

INTRODUCTION

Prior to 1987, information related to water policy was available in some policy statements made in the legislatures, as also in the working of public offices. In 1986, a National Water Resources Council (NWRC) chaired by the prime minister of India, was constituted. The body adopted a National Water Policy in 1987. Under the impact of changing water-related concerns, the NWRC adopted a revised water policy in 2002. A comparison of the features of these policies relevant to IWRM, is given as in Table 2.2.

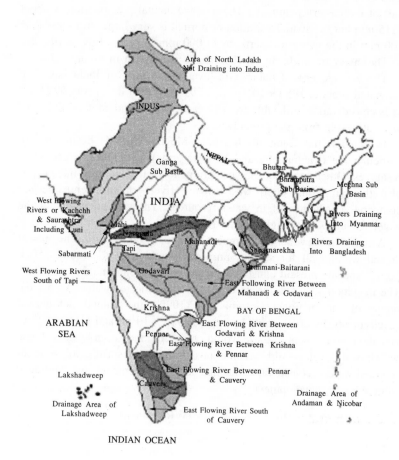

Figure 2.1: River basins of India

Table 2.1: Indian water resources—Spatial distribution

Basin/Basin ground	Percentage of country's area in	Percentage of national water resources
Brahmaputra and Meghna arms of the GBM basin	7.2	34.7
Ganga arm of the GBM basin	26.2	26.9
Subernarekha, Bramhini, Baitarni, and Mahanadi	5.9	10.5
Narmada	3.0	2.4
West-flowing river south of Tapi	3.4	10.3
All other river basins	54.3	15.2

The extent to which the policy has progressively accepted the principles of IWRM is discussed below.

PRINCIPLE I: WATER AS A FINITE AND VULNERABLE RESOURCE

The finite nature of water resources is being fully addressed in all water polices and water-related laws. Most of the interstate water-related discords and their solutions fully recognize this principle.

The vulnerability of the resource to pollution and its vulnerability in terms of low flow reduction and consequent effects on water-related ecosystems has now been addressed in the policies. In recent years, the Central Water Pollution Board laid down stringent standards about the quality of effluent to be discharged in natural waters. Policy guidelines and standards in regard to the quantum of low flow which needs to be maintained in the natural streams are however not yet laid down.

PRINCIPLE II: PARTICIPATORY APPROACH

This principle is mentioned in all policy documents but detailed policy and legal measures to ensure participation were not available earlier. Currently, four states namely Andhra Pradesh, Madhya Pradesh, Tamil Nadu, and Orissa have already passed laws to ensure participatory irrigation management and perhaps more states may follow. However, participatory management for other uses, as also for the basin as a whole is neither envisaged in any law, nor practised.

PRINCIPLE III: THE IMPORTANT ROLE OF WOMEN

Although the umbrella policy documents accept this principle, detailed strategies to achieve this are missing (see section on 'Integrating the gender-related concerns in water management').

Table 2.2: Comparison of Indian water policies

Topic	Salient features of the 1987 policy (Paragraph number of policy mentioned in parenthesis)	Salient features of the 2002 policy (Paragraph number of policy mentioned in parenthesis)
Need for the policy	Water is a scarce and precious national resource to be planned, developed, and conserved.... On an integrated and environmentally sound basis ... (1–8)	Added 'management' ... (1–4)
Water resources planning	Water will become scarce in future (1–8)	No change
		Emphasized the need to practise non conventional methods such as inter-basin transfer, artificial recharge of groundwater, and traditional practices like rainwater harvesting
Basin or sub-basin as a unit	Planning has to be done for a basin or a sub-basin. All individual projects to be considered within the framework of this plan (3.2)	Added 'management' aspects—'surface and groundwater', 'sustainable use', 'quantity and quality aspects'
Institutional mechanism		Emphasized reorientation or creation of institutions towards multi-sectoral, multi-disciplinary and participatory development and management, integrating quantity, quality and environmental aspects (4.1)
		Brought out the need for river basin organization for optimizing the use, having regard to agreement and awards under relevant laws. The scope and power of the organization however have been left to the states themselves.

(Contd.)

Table 2.2 Contd.

Topic	Salient features of the 2002 policy (Paragraph number of policy mentioned in parenthesis)	Salient features of the 1987 policy (Paragraph number of policy mentioned in parenthesis)
Priorities in use	Drinking, irrigation, hydropower, navigational and individual (8)	After 'hydropower', inserted 'ecology, agricultural and other industries, navigation and other uses' (5)
Project planning		Emphasized of the study of socio-economic and environmental impacts (6.2), sustainability (6.3). Projects in hilly areas (6.4) and need for separate economic evaluation of projects for social weak groups (6.3)
Stakeholders participation	Efforts need to be made to progressively involve farmers in irrigation management (12)	'Management of water resources for diverse uses should be done by adopting a participatory approach, by involving ... stakeholders ... in planning, designing, development and management. Necessary legal and institutional changes should be made, duly ensuring appropriate role for women.' Involvement of WUAs in municipalities, local bodies, panachayats and eventual transfer were emphasized. (12) Privatization possibilities have been brought out (13)
Water Quality		Issues like effluent treatment minimum flow maintenance are brought out, the 'polluter pays' principle prescribed and supporting legislation in water quality preservation is advocated (14).

PRINCIPLE IV: WATER AS AN ECONOMIC GOOD

Even while emphasizing both the social and economic nature of water, the polices do stress the need for increased water prices towards financial sustainability. However, the implementation of the policy in this regard has been rather difficult.

'Integration' in IWRM

INTEGRATION OF VARIOUS WATER USES

Irrigation and hydropower: Irrigation and hydropower uses have been integrated very well in the planning and management of projects built in the last fifty years. In the Indus river basin (Figure 2.2) (Bhakra-Nangal, Beas-Sultej diversion, Pong dam-Rajasthan Canal) a remarkable integration has been achieved so that all the water used in hydropower production gets used in irrigation as well. The main management tool is the Reservoir Regulation Committee of the Bhakra Management Board, in which the two interests and all states are represented, and which, by a tradition, has been able to take workable decisions without getting entangled in controversies.

A somewhat similar and good integration is available in the Chambal cascade of the Ganga sub-basin (See Figure 2.3) and many other projects. The Rihand-Sone Barrage complex in the Ganga arm represents a system where such integration between hydropower use in one state and irrigation use in another was not planned, but reasonably good results are obtained by having a reservoir regulation committee, chaired by an official of the Union Government, as an advisory body.

The Koyna project in the Krishna basin (Figure 2.4) represents a conflict of interests. Hydropower generation favoured a transbasin diversion of water to near sea level, across the ridge, thus making the water unavailable for irrigation. Since, in the basin, irrigation from these water was mainly in other states, the conflict of interest became an interstate issue, and this was solved in the award by restricting the hydropower diversions.

Flood Control and Conservation: Integrating use of water for conservation purposes with the objective of flood control was also found to be difficult, since the resource of potential storage capacity is severely restricted, by topography, settlement patterns, and high costs. The Hirakud reservoir in the Mahanadi basin (Figure 2.5) was planned as a remarkable compromise. Taking advantage of the cyclic changes in the flood frequency regime, much of the storage space was reserved for flood

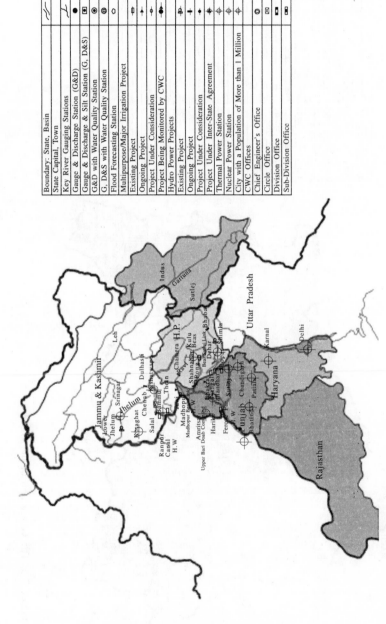

Boundary: State, Basin	
State Capital, Town	
Key River Gauging Stations	
Gauge & Discharge Station (G&D)	●
Gauge & Discharge & Silt Station (G, D&S)	▣
G&D with Water Quality Station	◉
G, D&S with Water Quality Station	⊕
Flood Forecasting Station	○
Multipurpose/Major Irrigation Project	
Existing Project	⬒
Ongoing Project	⬙
Project Under Consideration	◈
Project Being Monitored by CWC	◈
Hydro Power Projects	
Existing Project	⬓
Ongoing Project	⬧
Project Under Consideration	◆
Project Under Inter-State Agreement	⬦
Thermal Power Station	⬦
Nuclear Power Station	⬦
City with a Population of More than 1 Million	
CWC Offices	
Chief Engineer's Office	⊘
Circle Office	⊠
Division Office	◧
Sub-Division Office	▮

Figure 2.2: Indian part of the Indus basin

Boundary: State, Basin	
State Capital, Town	
Key River Gauging Stations	
Gauge & Discharge Station (G&D)	●
Gauge & Discharge & Silt Station (G, D&S)	▣
G&D with Water Quality Station	◉
G, D&S with Water Quality Station	⊚
Flood Forecasting Station	○
Multipurpose/Major Irrigation Project	
Existing Project	
Ongoing Project	
Project Under Consideration	
Project Being Monitored by CWC	
Hydro Power Projects	
Existing Project	
Ongoing Project	
Project Under Consideration	
Project Under Inter-State Agreement	
Thermal Power Station	
Nuclear Power Station	
City with a Population of More than 1 Million	
CWC Offices	⊗
Chief Engineer's Office	⊠
Circle Office	
Division Office	■
Sub-Division Office	

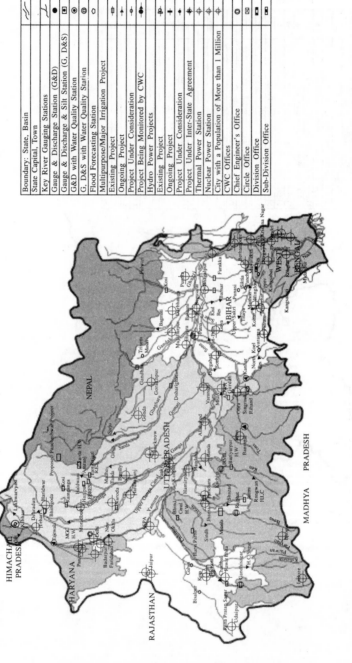

Figure 2.3: The Ganga arm of the GBM basin

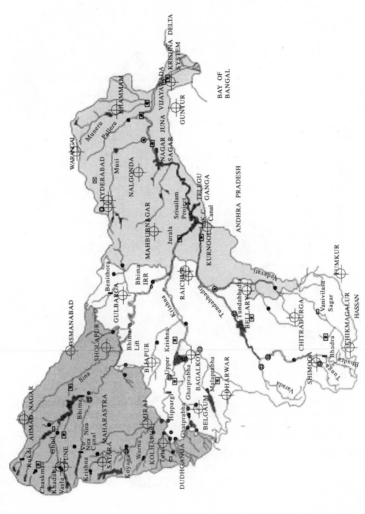

Figure 2.4: The Krishna basin

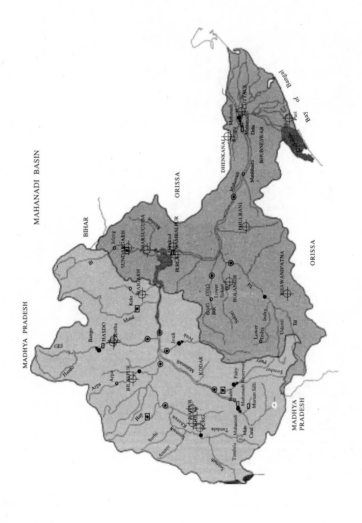

Figure 2.5: The Mahanadi basin

control in the beginning and middle part of the flood season, and was then gradually and fully transferred towards conservational uses by the end of the wet season. This worked remarkable well, until, due to its upstream development, the inflows reduced and to obtain a reasonable possibility of filling, the rule curves had to be frequently changed to allow an earlier filling. Thus, the objective of flood control had to be somewhat compromised, although the project continues to yield large benefits of flood control, irrigation, and hydropower.

The main Damodar valley reservoirs (Maithon and Panchet) in the Ganga sub-basin also tried integration of these conflicting uses. Flood control space was kept as a separate reservation, but the land in the top strip was not fully acquired. It was felt that in the event of large floods causing upstream submergence, the flood control benefits downstream would be far larger than the upstream damages, which could be compensated. In spite of many management innovations, this has not worked well, since it resulted both in an upstream-downstream conflict and in interstate issues. Perhaps, learning from such situations, the Rengali project on Bramhini and the Chandil–Ichha complex on the Subarnrekha were planned with a full and separate reservation of flood control space, with full land acquisition.

On the other hand, the planning of the Ukai multipurpose reservoir on the Tapi is an example of a lost opportunity in addressing the flood control problem of Surat through the project.

Irrigation and Domestic Water: Integration of irrigation with domestic/ municipal water supply again has been somewhat difficult. Much of these difficulties can be traced to a lack of foresight in projecting the fast growing municipal uses, and to a lack of institutional integration. The irrigation departments, even when converted into water resources departments did not internalize other concerns, and gave an overriding priority to protecting the interest of irrigation. Limiting irrigation areas in view of growing municipal demands, and reducing the water allowance for irrigation gradually, could have reduced the distress to irrigation in accommodating higher municipal requirements, but this has often not been done. The allocation of the more reliable part of available water for municipal use, as also reserving the insufficient water in drought situation for municipal purposes, has also been tried, but some conflicts have developed. For example, in Gujarat, the Sabarmati basin farmers were agitated over higher urban use priorities, and in the Rajkot area, full reservation of residual reservoir water, during a drought situation, almost caused a riot. Similarly, large shifts in allocation of waters from irrigation to municipal use has caused widespread tensions in Maharashtra. The higher reliability for municipal use is reflected in the priority given for Delhi's use in the interstate upper Yamuna accord; but implementing this had its share of difficulties.

In the author's view, the highest priority accorded to the domestic use of water, in the national policy, needs to be elaborated in other supporting documents, to reflect priorities in quantitative allocation, priorities in allocating the best quality available water for this use, and a priority in giving the most reliable part of the stochastic resource, for this purpose. Such an elaboration has not been done as yet.

INTEGRATING THE UPSTREAM AND DOWNSTREAM INTERESTS

Introduction: In general, water projects for irrigation or flood control benefit areas downstream of the diversion on storage structures, and could cause some damage through submergence, or through restraints on future use, on the upstream. Similarly, upstream project in a basin can reduce availabilities downstream.

Watershed management: Watershed management, involving a wise use of land and water, has to be an integral part of IWRM. Often, this is the only opportunity available to the upland hilly and sloping areas in regard to water management. In regions with low rainfall, watershed management, in conjunction with other means of irrigation, can use the local rainfall before tapping other costlier sources. In India, watershed management has been practised at many places, but there are hardly any instances of its integration in the overall development plan. Also, rainwater harvesting in irrigated fields, by storing the otherwise non-effective part of the rainfall in field ponds has not been practised. Further, good success stories like Ralegaon Shindi and Adgaon (Maharashtra), the Alwar region (Rajasthan), etc., have been considered as separate alternatives and not as a complementing component of a master plan. Finally, in regard to soil conservation, the objective of reduction of sediment load and consequent improvements in the reservoir efficacy has been overemphasized, and the aspects of soil and water conservation have been underempahasized, perhaps with limited institutional objectives of establishing links with the available project finance. The resultant controversies about financing have impeded both the watershed management and the reservoir projects.

Irrigation benefits, upstream and downstream of reservoir: A pre-occupation with gravity irrigation, ignoring the widespread availability of energy, leads to an irrigation planning where water is carried, in a large project, over hundreds of kilometres to irrigate 'commanded' lands, ignoring the socially and economically superior claims of lands a few metres higher than the reservoir or head reaches of canals. This has also been adversely commented upon by the National Commission for Integrated Water Development.

Thus, in the vicinity of the Tungabhadra reservoir in the Krishna basin, unauthorized pumping for irrigation (which according to the author was an efficient use), developed to such an extent that a cognizance of this had to be taken in the interstate board monitoring the allocations. In the same basin, such anomalous situation was sought to be corrected in planning the Ujjani reservoir (Maharashtra), where such lifts were planned from the beginning.

Another way of integrating the upstream aspirations is to provide for these in planning the downstream developments. This has been done notably in the Bansagar interstate agreement on the Sone river (Ganga sub-basin). But a failure to project such unavoidable and uncontrollable uses resulted in water availability problems for the Hirakud reservoir in the Mahanadi and for the Gandhisagar resources on the Chambal river (Ganga sub-basin).

INTEGRATING THE GENDER-RELATED CONCERNS IN WATER MANAGEMENT

Very little is being done in this regard, perhaps because the concerns and the possible solutions are not known. Watershed management, afforestation, and resettlement are obvious areas where addressing such concerns appears essential. The high priority which the policies give for domestic water use can be put into practice if women were empowered in decision making. The option to practise night irrigation could be another such area.

Changes in legal provisions to allow and encourage women family members to be elected to Participatory Irrigation Management (PIM) committees even if they are not the owners, perhaps need to be considered. This has been done in Orissa. The law concerning participatory irrigation management in Orissa allows landowners to nominate another family member as a member of the committee, and this in turn has facilitated the participation of women, and also the establishment of an all-women irrigation management society for one area. Creation of a separate women's cadres in irrigation staff can perhaps give a greater impetus to the gender-related issues in practising irrigation. This has not been tried as yet.

INTEGRATING THE INTEREST OF ETHNIC MINORITIES AND SOCIALLY DEPRIVED GROUPS

In areas with large tribal population, the economic feasibility criteria have been made less stringent (reduced cut off B:C ratio), and special care for financing and monitoring is being taken by the government through a separate plan. For other deprived sections, the general philosophy is that a large overall development would automatically benefit these groups. In regard to reservoir oustees, which are perhaps more likely to be tribals, separate and stringent monitoring of the resettlement plans is being undertaken. However, there seems to be a need for generating more information about the benefits reaching these groups and then for developing innovative methods for improving the situation.

INTEGRATING THE AFFECTED AND THE BENEFITED IN A SINGLE GROUP OF BENEFICIARIES

Although much progress, both in developing appropriate policies and in improving procedures has been achieved, cases like Sardar Sarovar (Narmada, Gujarat), and Tehri (Ganga, Uttaranchal) resulting in apparent dissatisfaction have become well known. In recent times, large and complicated rehabilitation and resettlement for the Almatti (Krishna Basin, Karnataka), Rajghat (Betwa-Ganga basin, Madhya Pradesh and Uttar Pradesh); Bansagar (Sone-Ganga basin, Madhya Pradesh), Ranjit Sagar (Ravi-Indus, Punjab, Himachal, and J&K) have been completed with comparatively much lesser tensions as a result of improved packages and responsive field management.

Higher cost of the rehabilitation and resettlement package, and not the lack of willingness, seems to be a constraint. Another constraint is the tardy implementation of well-meaning policies, leading to non-settlement of small grievances, which in turn leads to large-scale dissatisfaction.

INTEGRATING THE DIFFERENT PHASES OF THE HYDROLOGIC CYCLE

Precipitation and run-off: The improved in situ use of precipitation, by ponding it in situ on the land surface, in the root zone or in the groundwater, is an integral part of watershed management.

Surface and Groundwater: The conjunctive use of the surface and groundwater is an important complement of the water policy in India, and detailed guidelines for such a use in surface irrigation projects have been prepared (CWC, 1997). According to the author's classification, the conjunctive use of the two sources can involve:

1. The use of groundwater storage space to modify the temporal availability of the water resources;
2. The use of groundwater aquifers to transport surface water;
3. The use of surface distribution system to transport groundwater from water-rich to water-poor area;

4. The use of groundwater aquifer as a filter to improve surface water quality; and
5. Mixing of surface and groundwater to improve the suitability for irrigation use.

Of these, in India, very good examples of the type (1) have been developed, mostly by the initiative of individual farmers, in the low rainfall alluvial tracts. The Kharif channels (i.e. channels flowing only in the wet season) of Uttar Pradesh (Madhya Ganga and Eastern Ganga) encourage large use of surface water in the high flow season for Kharif crops including paddy. In the long dry season (winter and spring), the canals are closed, but the recharged groundwater is pumped out for irrigation. Even in other perennial surface systems in these regions, shortages in the dry period are made good through groundwater pumping.

In Figure 2.6, for many districts of Uttar Pradesh, the comparison between the artificial (irrigation-induced) groundwater recharge, and the groundwater use, both in terms of the ratios of these quantities with the natural recharge is depicted. The figure clearly brings out how the use of groundwater increased with the availability of irrigation-induced recharge.

Also, in the brackish groundwater area of the western alluvial tracts (Indus basin, Punjab and Haryana) there are examples of mixing of the two

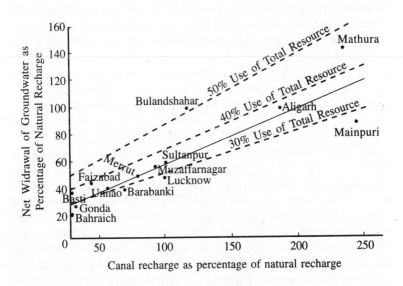

Figure 2.6: Conjunctive use in districts of Uttar Pradesh

waters. Similarly, in the Yamuna area of Haryana, the plentiful groundwater locally available near the river is transported through the surface canal network and used elsewhere.

INTEGRATING STAKEHOLDERS

India traditionally had a mixed system for institutions regarding water development and management. Large systems such as the Cauvery delta irrigation of the Cholas, the Vijaynagar canals (fifteenth century), etc., were built and managed by the state; but even in such systems, lower-level irrigation management was privatized (the satta system in Sone, Bihar). Some private Zamindari canals (eighteenth and nineteenth century) were built. Small irrigation systems however had traditional cooperative systems, such as the water harvesting tanks of Paliwals in Rajasthan, the Phad system on tanks in Maharashtra, and the Kuddimaramat in Tamil Nadu. With the establishment of the British rule, the state usurped the ownership and management of waters. Although the Phad and Kudimaramat have survived, in general the surface water use was almost without any stakeholder participation.

In the last five years, large changes in policies to favour PIM have been made. However, in general the progress in regard to PIM is very slow, and this could be due to the non-resilience of the well-established irrigation departments. The higher-level task of devolving basin management responsibilities to a body of stakeholders has not taken off at all in spite of strong recommendation of the National Commission. This could be due to a feeling that the government, in a democracy, represents all stakeholders and is the largest NGO. Since most basins in India are interstate, the discords among the different state governments seem to deafen the separate voices of stakeholders within the states.

INTEGRATING THE GOVERNMENTS: POLITICAL ASPECTS OF IWRM

International Waters: India is a co-basin state to the Indus basin, the Ganga-Brahmaputra-Meghna basin, the Irrawady basin, and some small streams.

In regard to the Indus (Figure 2.2), the Indus Water Treaty with Pakistan has stood the test of time through wars and hostilities. It is often cited as a very good example of international cooperation in waters. However, it is hardly an example of IWRM. It is bilateral and not basin-wide (China and Afghanistan are basin states which are not party to the treaty). Also, in its structure, it divides and shares the river of the Indus basin by allowing independent developments rather than integrating the development. As one example, the Upper Bari Doab canals, with headwork in India, had to cut off the irrigation to some tail areas in Pakistan for which separate provisions had to be made. One doubts if in the given circumstances, any arrangements based on IWRM principles, even if providing larger benefits to both the parties, could have become practicable. However, future efforts towards better integrated development may have to be initiated.

The Ganges treaty with Bangladesh is again a bilateral treaty which regulates the releases made downstream of the Farakka barrage. It, however, has some integrating elements in regard to the need of augmentation of supplies and in regard to maintenance of the river regime. In comparison, the Mahakali treaty with Nepal is even more integrating, in as much as it aims at a joint use of the contemplated facilities by both the nations towards an optimum development.

Interstate Waters: About 92 per cent of India's territory is drained through interstate (including international) basins, and thus most schemes for water development could have an interstate angle. As development proceeds and water stress develops, interstate issues become prominent.

India has a semi-federal structure, and both the Central (Union) and the state governments have executive and legislative responsibilities regarding water development and management. The states are the main players, even in interstate basins, unless the Union legislature passes an act, defining the role of the Union in a general or specific way.

The interstate disputes within India, such as those about Cauvery, Narmada, Ravi-Beas, Godavari, and Krishna are well known. What is less known is that a very large number of discords have been well managed, and a very large number of interstate agreements have been signed among various states (see Central Water Commission—Legal instruments, 1997). Mohile and Illangovan (1995) studied the implications of this semi-federal set-up, which does not allow a single-point decision making about integrated basin developments on the Indian water development scene. They conclude that although the resultant developments are sometimes sub-optimum and ignore better-looking alternatives requiring interstate cooperation at project levels, no serious effects or lopsided development have resulted.

Even then, interstate disputes, although unavoidable, are serious irritants, causing delays, and improved institutional arrangements appear necessary. These are brought out in the concluding remarks.

A brief basin-wise review of some of the interstate problems is given below:

Krishna Basin (Figure 2.4): The tribunal's award was the basis of the settlement. In the award, quantitative allocations and restriction on use, diversions, etc., are included. What is missing is a scheme for implementing large interstate projects which could, for example, store water in one state and utilize it in other states. Thus, a very workable, but sub-optimum, framework became available through the award. However, as the stress grew, and the period after which the award could be reopened for a review neared, subsidiary disputes about the interpretation of the award came up (for example, discords about Telugu Ganga, Almatti, etc.).

Godavari basin (Figure 2.7): The award was largely based upon interstate agreements reached outside the tribunal. The award is eminently workable, and no subsidiary disputes have developed. However, somewhere on the lines of the Indus treaty, the arrangement is based more on allocating sub-basins/basin parts statewise than on integrating the developments.

Narmada basin (Figure 2.8): The core of the award and the resulting development plan is the creation of two large interstate reservoirs, with shared costs, benefits, and impairments, to be operated in an integrated way. Although a large number of developments within the state are also provided for, the Narmada can be sited as the best example of integrated development resulting from an adjudicated settlement, defying the general observation that adjudicated settlements tend to miss the optimum based on cooperation and thus tend to be inferior. However, this settlement has later come under a considerable stress, both in regard to its interpretation, and through proposals for making changes in the plans through mutual agreements envisaged in the award.

Upper Yamuna: The Upper Yamuna sub-basin discord was settled through an agreement amongst the five states, brokered by the Union Government. However, subsidiary disagreements have developed on interpretation. Also, the states which devolved their regulatory functions of the main system to a Board created by them through the agreement were not perhaps prepared to accept such a devolution. According to the author, a likely failure of the accord was avoided due to the overriding directions of the judiciary.

Sone: The Sone sub-basin agreement was similarly reached in 1973 among the three states and it paved the way for the construction of the interstate Bansagar reservoir project. The agreement came under some stress since a review of the water availability through the Sone river commission, as envisaged under the agreement, was not unanimously accepted. Also, the agreement seems to have allocated all the waters for irrigation use, by ignoring other uses. However, other provisions, including implementation of the Bansagar reservoir project, have stood well.

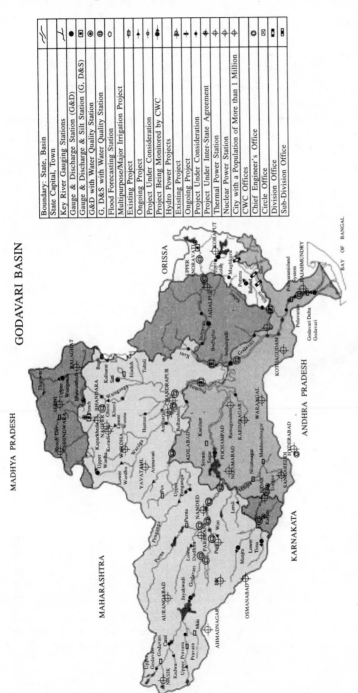

Figure 2.7: The Godavari basin

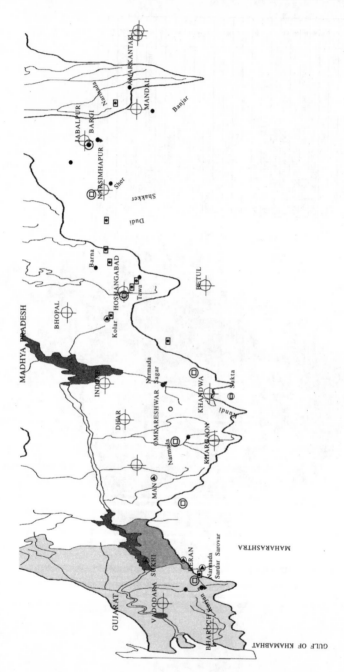

Figure 2.8: The Narmada basin

The Sone and Upper Yamuna agreements, according to the author, open another question which is not fully debated. Can co-sub-basin states reach an agreement among themselves to allocate the water of a sub-basin, without the agreement of other co-basin states which are not members of the sub-basin? Although sub-basin agreements may be practicable and pragmatic, they miss the opportunity of integration amongst sub-basins. For example, can it be presumed that both the Upper Yamuna and the Sone have little responsibility towards maintenance of an acceptable low flow regime in the Ganga? These questions need to be debated. Perhaps the Union should have a responsibility of analysing and approving the interstate agreement in interstate river basins, and a concurrent responsibility of ensuring the implementation of the approved agreements. The legal system in the United States of America seems to give such responsibilities to the federal government.

IWRM in the Basins of India

We have already reviewed the status of IWRM in India, and have illustrated this with various cases. Now, we assemble from this, the general status of IWRM-related issues in some of the basins, more often cited in different context. This is given in Table 2.3.

Concluding Remarks

The review brings out that a large experience about the various types of integration envisaged in IWRM has been gathered in India, but there is no 'model' or showcase of IWRM available. Integration in bits and parts or a 'flocculation' of sorts is evident, but the picture of a well-managed basin, with an integrated plan of development by various governments, other stakeholders, water use interest and so on, to the overall satisfaction of everyone, appears to be almost utopian.

The main difficulties do not appear to be in the various policy documents, all of which, explicitly or implicitly, favour IWRM. The difficulties seem to be in the legal and institutional systems and in the attitudes of decision makers in the governments. The decision makers in India, time and again, seem to treat water as a property to be allocated and owned.

Difficulties also result from the semi-federal set-up. In general, the current provisions in the Constitution and the laws for regulating the water-related responsibilities through a semi-federal set-up seem to have worked fairly well earlier, although these are coming under considerable

Table 2.3: Basin-wise position regarding IWRM

Basin/Sub-basin	Priority Item	Remarks
Krishna basin	Koyna	Conflict of hydropower and irrigation resolved through restrictions on hydropower diversions
	Ralegaon Shindi	Good watershed management not integrated in basin plans
	Tungabhadra	Reservoir lifts, not planned but long gravity transfers implemented leading to unauthorised use
	Ujjani	Reservoirs lifts integrated in planning, with downstream irrigation
	Almatti	Smooth integration of impaired and benefited
	General	Very workable tribunal award, but this misses the optimum and does not integrate the water development across states. The settlement has come under stress and many subsidiary discords have developed
Godavari		Very workable award, without development of subsidiary discords, but not fitting in IWRM concepts
Narmada	Sardar Sarovar	Dissatisfaction about integration of the impaired group with benefited. Adjudicated settlement, but yet provides for an integrated development plan. An outstanding example. However, the award has come under considerable stress, and a desirability of future agreements overriding the award have been advocated by some
Mahanadi	Hirakud	Joint use of capacity, integrating flood control and conservation. Difficulties with upstream use, resulting in some dilution of flood control objectives
		Non-integration of future upstream uses with downstream leads to water availability problems

(Contd.)

Table 2.3 Contd.

Basin/Sub-basin	Priority Item	Remarks
Indus system	Bhakra-Beas	Remarkable integration of irrigation and power regulation committee
	Ranjit Sagar	Smooth integrating of the impaired and the benefited
	General	Good example of international agreement but hardly an example of IWRM
Chambal sub-basin (GBM)	Chambal Cascade Gandhisagar	Good integration of hydroelectric and irrigation interests / Non-integration of future upstream uses with downstream irrigation in planning is causing water availability problems
Upper Yamuna (GBM)	Yamuna accord	Higher priority accorded to water supply but management difficulties experienced
	Western Yamuna Canals	Integration of surface and groundwater through transport of groundwater
	General	Discord settled through interstate agreement. Serious management difficulties averted by the directions of the judiciary. Appropriateness of sub-basin agreements in question
Upper Ganga (GBM)	Tehri	Dissatisfaction about integrating the impaired group with the benefited
	Madhya Ganga and Eastern Ganga	Remarkable integration of surface and groundwater through conjunctive use
	Rihand	Lack of integration between hydropower and irrigation in planning, partly corrected in management
Sone sub-basin (GBM)	Bansagar	Smooth integration of the impaired and the benefited
	Sone-Bansagar agreement	Interstate agreement paved way for implementing a key storage. However, the agreement has come under a stress. Also, the appropriateness of a sub-basin agreement in question
	Sone Canals	Traditional privatization of field management (Satta)

stress now. Perhaps the situation can be improved by the following measures:

(a) The Union Legislature could pass an act which requires that the broad planning and management aspects of the schemes to be implemented in interstate basins require an approval by the Central government after due consultation with other co-basin states.

(b) The Union Legislature could pass an act to empower an agency to plan and implement schemes for inter-basin transfer of water to non-co-basin states, after due consultations with co-basin states.

(c) Basin-wise authorities to oversee the development and management of interstate basins are set-up, with the participation of all stakeholders, including the state governments.

(d) Policy guidelines about allocation of water of an interstate river basin are finalized and notified under the relevant act; and these lay down the general principles of allocation, and the machinery to oversee the management in accordance with the allocation.

(e) A system of approving and recording interstate agreement is introduced by the Centre.

(f) A well-documented regime of water rights is established, which would, inter alia, define the rights of individuals, landless and landholders, communities, and states. It would bring out the dynamic nature of the rights and the need for periodic reviews of the rights, and would discuss the transferability and tradability of these rights.

Some difficulties in implementing IWRM can also be traced to the fragmented institutional structure in the public domain concerning water, and a poorly developed structure in the non-governmental sector.

The spatial limits of integration have also to be understood in the process of implementing IWRM. As noted by GWP, sovereignty requires a special conflict-related mechanism, and the mechanisms implemented at the national levels do not automatically translate into validity at the international level because of the well-established international principal of national sovereignty.

One wonders if this principle has a corollary in regard to the semi-federal set-up as found in India. The Indian states, by no means, have sovereignty; but they have a large measure of autonomy. In addition, many states have a subculture or almost a sub-nationality based on languages and other considerations. The national principle of maintaining such an autonomy could seem to make some of the normal mechanism for conflict resolution loosen their validity in regard its interstate matters.

The better workability of the solutions to the conflicts in the Indus and Godavari, both of which are of the non-integrating type, needs to be considered in understanding the limits of the IWRM principles.

IWRM has evolved from the earlier concepts of IRBM (integrated river basin management). In the present form, IWRM lays considerable stress on the basin being a unit of planning and management even while recognizing the conflict between these concepts and the national sovereignty. The basin concept itself depends on the basin being a 'hydrologic unit', by its having a 'common terminus' and on the interdependence of actions within, and possibilities of 'integration' within this unit. The author has advocated that as problems become more complex, the common 'terminus' beyond which the considerations end can shift downstream. The inter-basin integration, possible and seemingly necessary in India (Ministry of Water Resources, National Perspective, 1980), through the plan of long-distance water transfers and other such proposals, would require a new look at the basin concept and a new interpretation of IWRM.

Thus, even while IWRM principles have not been fully applied in India, we are already looking forward not only to their fuller application, but also to their dynamic interpretation. IWRM is likely to be a desirable and changing target to guide water management in India for quite some time. This is as it should be, since IWRM is a process.

References

'Guidelines for planning conjunctive use of surface and groundwater in Irrigation Projects', Central Water Commission, 1995.

'Integrated Water Resources Development: A Plan for Action', Report of the National Commission for Integrated Water Resources Development', Ministry of Water Resources, Government of India, 1999.

'Integrated Water Resources Management', TAC Background Paper No. 4, GWP, 2002.

'Legal instruments on Rivers in India', Volumes I to IV, Central Water Commission, 1997.

Mohile, A.D., and M. Illangovan, 'Indian experience in River Basin Planning and Management', Indo-French Seminar on River Basin Management, New Delhi, 1995.

'The National Perspective of Water Resources Development 1980', Ministry of Water Resources, Government of India.

'The National Water Policy 1987', Ministry of Water Resources, Government of India.

'The National Water Policy 2002', Ministry of Water Resources, Government of India.

3

SABARMATI RIVER BASIN (INDIA): PROBLEMS AND PROSPECTS FOR INTEGRATED WATER RESOURCES MANAGEMENT

C.D. Thatte

Integration and Sustainability

IWRM TO IWRDM

After a burst of activities related to water resources development in the middle of the twentieth century, the world community became complacent, relegating the 'freshwater' issues to the backburner of global economic agenda. The subject got a shot in the arm again with the recognition accorded to it in the 1992 Earth Summit at Rio, followed by the World Water Fora (WWF) held in 1997, 2000 and 2003. The Agenda 21 of Rio for 'Sustainable Development', under Chapter 18, dealt with the subject a little, but recognized 'Integrated Water Resources Development and Management (IWRDM)' as the key issue. Some time down the line thereafter, the developed world, which had accomplished water resources development by then, considered it redundant and coolly dropped it, shortening the 'IWRDM' to 'IWRM'. No eyebrows were raised, nor reasons given. Also, the concept of 'integration (I)' in the acronym, remained unelaborated.

The WWFs, the World Food Summit (WFS, 1996), the Millennium Summit (2000), the World Summit on Sustainable Development (WSSD, 2002), and the World Food Summit, Five Years later (WFS + fyl, 2000), all however reiterated, one way or the other, the criticality of the subject by linking it to an action plan to halve the thirst (drinking water), hunger (water for food), and consequent poverty and ill-health by 2015. The WSSD identified WEHAB—standing for water, energy, health, agriculture, and biodiversity—as the critical global agenda. All the five components are undoubtedly closely linked with IWRDM. The WSSD also highlighted the need to forge community participation and partnerships amongst stakeholders to promote the agenda. Special attention to empowerment of weaker sections of society represented by women, backward classes, tribals, and minorities was reiterated.

The last summit also acknowledged the failure of global community during the last five years, in attaining these targets. As a grim reminder, presently twenty-one countries of the African continent face famine conditions and insecurity with respect to food, drinking water, energy, fibre, and fodder, exacerbating the poverty and deprivation. Contrast this situation with developing countries of Asia such as India, where, in spite of the century's worst drought during 2002, thanks to the green revolution the food stocks are at their highest, the country is becoming the biggest exporter of wheat and rice in Asia, inflation is at about 3 per cent, foreign exchange reserves are at an all-time peak, and there is political stability and democratic strength. Countries such as India, however, cannot afford to be complacent, as population keeps growing, some sections of people are still food insecure, and several non-structural measures are yet to be implemented for achieving harmony and integration.

A recap of the issues therefore will not be out of place. Annual freshwater availability around the globe, barring a few fortunate geographic regions, varies widely in time and space, necessitating its 'development (D)' and then its deployment through appropriate 'management (M)' for serving beneficial needs of the mankind. The largest component of global use at about 70 per cent of the total freshwater abstracted is for agriculture, which produces food for human consumption and fodder for rearing cattle to get meat. The component goes up to 90 per cent in several developing countries. It will continue to remain largest component, simply because production of one kilogram of food consumes about one cubic metre of water, whereas daily metabolic requirement of a human body is only four litres. Human needs other than food, like drinking water, are non-consumptive in nature and will continue to remain relatively small. The ecosystems use up a large component right from the precipitation stage but the quantum so used remains mostly unquantified.

The numbers of days in a year when precipitation takes place as also its intensity varies from region to region, resulting in droughts and floods which constitute the extreme manifestations of the hydrologic cycle. Situations, of course, vary widely in temperate, humid, semi-arid, arid, or tropical, mediterranean, monsoon climates. At places, the year's rainfall occurs in a couple of storms lasting for a few days covering less than 100 hours. The proportion of minimum: average: maximum annual rainfall is 1:10:100. Run-off generated due to such storms has to be captured behind storages for use round the year. In those fortunate regions where the distribution of rainfall is well spread round the year, people do not require such development. The human population keeps growing, although, freshwater occurrence is constant. The water availability and the water users both are unevenly distributed over different regions of the world. Prudent 'development and management' of freshwater alone helps mankind cope with natural variability besides meeting with human and ecosystem needs, and as such should be on the world's agenda related to society's well-being, economics, and related policy.

A look at the systems of water management for agriculture is indeed interesting, not only because agriculture accounts for the largest use of water, but also because being the oldest critical sector, data are readily available. Arable areas round the world either face shortage or excess of required moisture regime during the crop-growing seasons. As is well known, irrigation from surface or groundwater sources and drainage systems (I&D Systems) respectively, remedy the situation. Still it is not well known that about 70 per cent of world's arable lands are not yet provided with either of these I&D Systems. It is not that these arable lands do not need them. The fact is the world has ignored them or in some cases vested interests have lobbied against such interventions. As a result, in some of these areas, poverty, undernourishment, ill-health, and deprivation is rampant.

The reasons for this lack of interest include lack of adequate and appropriate governance, policy, strategy, priority, lack of resources, wastage, inefficiency, lack of finance, lack of export earning to enable import of virtual water, economic poverty, lack of institutional capacity, overseas agricultural interests, subsidies, adverse trade regime, local strife, dictatorships, military regimes, tribal warfare, terrorism, wars, remnant colonial pressures, and others. Table 3.1 indicates continent-wise status for the ninety-six countries in the International Commission on Irrigation and Drainage (ICID) network, which represents almost the whole world scenario. The reasons for the imbalance and the insecurity in different continents can be visualized from the proportion of areas not served by any of the I&D systems. The reasons become clearer when read with the proportion of undeveloped water resources potential for each continent, as roughly estimated from different sources.

The table indicates that Europe, America, and Australia have more or less developed their water resources potential. Water resources

Table 3.1: Status of water management and lack of water development in the world

Continent	Population	Geographical Area	Arable land		Water (M.ha.)		Management		Practice		Undeveloped potential
	Millions	M.ha.	M.ha.	% (5/4)	Irrigation	% (7/5)	Drainage	% (9/5)	Nil	% (11/5)	%
Asia	3508	2745	531	19	178	34	49	9	303	57	30–50
America	757	3771	370	10	40	11	65	18	266	72	10
Europe	703	2202	295	13	27	9	49	17	219	74	10
Africa	598	1820	157	9	11	7	4	0	142	90	90
Oceania	23	803	56	7	3	5	2	0	51	91	10
Total	5588	11,341	1409	12	259	18	168	12	982	70	40
World	5978	13,387	1512	11	271	18	190	13	1051	70	40

Source: ICID Database, FAO Statistics, IWMI, IFPRI (1999–2000).

development in Asia started around the 1950s and stands at about 60 per cent of the potential, whereas Africa is at the bottom of the ladder with less than 10 per cent of its potential developed so far. The proportion of arable area without I&D systems in Africa to Asia is 60 per cent. The proportion for population is 17 per cent, but the geographical area is 66 per cent and arable area is 30 per cent. With the level of implementation of IWRDM, food sufficiency (albeit with local problems) is manifest in the Asian continent. On the other hand, need for water resources development systems in the African continent becomes apparent from the figures. Continued deficiency in this respect, no doubt, will be reflected in similar crisis cropping up now and then. No doubt, evolution of a strategy for going back to IWRDM from IWRM for them at least is obvious.

DEFINITION AND SCOPE

The Global Water Partnership (GWP) defines IWRM and includes 'development' in the term management as:

... a process which promotes the coordinated development and management of water, land and related resources, in order to maximise the resultant economic and social welfare in an equitable manner without compromising the sustainability of vital ecosystems.

How GWP in reality promotes development of water resources is another matter. The global practice mostly ignores or sidesteps the need for 'D' and coordinated implementation of 'D&M'. ICID has recognized in its Strategy for Implementation of Action Points emerging from the Vision for Food and Rural Development, the need for IWRDM. India happens to be one nation which categorically recognizes IWRDM in its National Water Policy of 1987 revised in 2002. Also, India's National Commission on development of an Integrated Plan for Water Resources recognized it in its report of September 1999, and brought out the need for IWRDM not only within each of the river basins but between the basins as well, to address cross-sectoral needs and demands. The Indian law and judiciary also have recognized this need as can be seen from pronouncements on several river water-sharing disputes between the states of the country. As the case study presented in this chapter is a report on the ongoing assessment of IWRDM work in a river basin of India, a revisit to the concept of IWRDM follows.

By the word 'integration' of water resources, one perceives a combination of several separate aspects into a 'whole' to achieve the identified goal of harmonized, sustainable, resilient, and flexible development pattern for humankind. The scope for 'integration' for freshwater, no doubt varies according to perceptions of different groups of end-users and availability of freshwater with all its variability. As it annually appears as a renewable resource through the hydrological cycle, circumscribed on the earth's surface by geography, a river basin is a logical natural resource unit. It permits correct hydrological assessments of water occurrence, its accumulation, transport by flow over or underground, interdependence of surface and groundwater, its consumptive and non-consumptive use, return flow and its reuse, basin efficiency, and water accounting. Integration therefore connotes upstream and downstream as well as catchment and or command areas. A river basin captures trade-offs amongst different uses and opportunity costs to optimize investments for maximizing production of goods and services. The basin connotes the boundary of the catchment area or basin drainage area incorporating all the land resources, including delta and coastal area for care, conservation, treatment, prudent use, regeneration, and reclamation.

In rocky areas, both surface and groundwater resource basins are largely congruent. In an alluvial part of a river basin, however, boundaries of different surface water basins become hazy, especially when the river enters a delta. The groundwater basins of two rivers often merge and do not match with surface boundaries and defy separate identification or labelling. Still, availability of both resources being dependent on precipitation on the river basin, the quantum is assessed accurately to the extent possible and considered together as a renewable dynamic freshwater resource to plan its development and management.

Some prefer to call IWRDM as IRBM denoting integrated river basin management, some as simply as IWM for integrated water management, or IRM as integrated resource management. Some also refer to CWSM denoting community-oriented watershed management. Integration basically therefore connotes—basin-wide optimized deployment for both surface and groundwater components through 'basin' level institutions for putting in place appropriate facilities, both off-river or in-river. Although, often some sections of society corner it to satiate their greed, prudence demands that the resource is treated as a national asset enabling its equitable sharing without causing any appreciable or significant harm to co-riparians in shared river basins. It is wrong to limit and equate IWRDM to basin boundaries. Ultimately, an administrative unit has to prevail for meeting demands.

IWRDM must aim to cover the basic minimum needs within a basin, leaving a designated agreed part for export to needy areas across the basin boundary, leading to inter-basin transfers. Supply and demand management are the keys for integration. However, it is unethical to preach demand management to those whose basic needs are not met with. Often conflicts arise when interests of societies aspiring for a share of the resource do not match. Management of such water-sharing conflicts at basin level and beyond basin boundaries needs special integrating efforts. As IWRDM is long gestating, planning has to begin in advance by a couple of decades. The period depends upon the institutional set-up in a country, state of its economy, and capacity of its people to take a comprehensive view. Presently, as the growth rate of population is showing a trend towards stabilizing by, say, AD 2060, it is possible to indeed take a long-term view about IWRDM for the ultimate size of population whose needs have to be addressed.

QUANTITY AND QUALITY

Human systems and ecosystems are two sides of the same coin viz. existence of life on the planet Earth and the two are best represented by the terms economy and ecology—both derived from the Greek word 'Eikos' indicating their interdependence. 'Eikos' means home—our planet earth, and the two terms convey the meaning of management and science of home respectively. The former has to defend and conserve the latter while the latter provides goods and services for the former. If it does, the human system is encouraged to defend and conserve the ecosystem. IWRDM has to address all facets of both systems, the former being at the centre of the latter covering flora, fauna, and biodiversity. But that interpretation calls for the full integration of the water resources sector in the economic system. Both consumptive and non-consumptive socio-economic needs have to be met with while ensuring social equity, economic efficiency, and ecological sustainability. These aspects have to be woven into the national policies and trade regimes and for that purpose stakeholder involvement is to be ensured at all managerial levels.

Freshwater availability depends upon climate, which is influenced by the synergy between the two systems. As water is a universal solvent, its quality changes in its organic and inorganic content as it moves downstream overland and through the soils and rocks. The natural processes of life forms also change it. The velocity of flow enables it to carry sediment picked up from the earth's surface. The river flow causes aeration, absorption of oxygen, and imparts a certain oxidation (purifying) capacity to the river water. The dissolved and suspended ingredients are released and absorbed, with chemical and physical processes causing a continuous process of change of quality of freshwater up to a certain limit known as its carrying capacity. Often, the capacity is exceeded by, for example, the biochemical-oxygen-demand (BOD) of water (which is a good indicator of proportion of organic pollutants of water due to the pollutants), till it is restored. Human systems can modify the quality more visibly and at a faster than natural pace—improve it for beneficial uses and/or further degrade it by releasing human wastes into it. These phenomena call for integration of processes of regeneration of such ecosystems if the carrying capacity is exceeded, to ensure availability of freshwater downstream in requisite quantity and quality. Starting from the catchment boundary, down in the river system and farther in the delta, estuary, and the coastal wetlands, the process of regeneration is important till the waters meet the oceans. While doing this, limits to the capacity of would-be beneficiaries from water resources development, to carry poverty and to meet their minimum needs, have also to be kept in mind. Development effort has to be in balance with regeneration effort. The predominance of the former cannot be wished away.

In pristine conditions, the balance developed between the availability of water, its flow and the ecosystem, does fluctuate with year to year climatic variation, manifest in particular due to floods and droughts. The balance gets modified with human interventions like river training, flood embankments, river jacketing or river-front development, flood zoning, use of flood plains for human activities, general flood control and management, storage, diversion, abstraction and return of waste, regeneration of water with a load of organic and inorganic chemicals. In case of thermal power stations, the temperature of released water being higher than the river water could cause adverse effect on the quality of water. Infrastructure development normally raises barriers affecting free movement of fish and fauna along and across river system, causing prevention of migration, and the loss of breeding grounds, spawning, and regenerative ability. Natural erosion and sedimentation processes and a seeming balance between them also are modified with these interventions.

Due to the nature and composition of human interventions, modifications are caused to both quality and quantity of goods and services offered by the ecosystems. Wastewater has to be treated before it is released back into the river system. Treatment should bring the water quality as far back to natural level as possible. But there are economic limits to treatment. BOD desired in a river system, for example, is less then 3 mg/l. The cost of treatment seemingly mounts steeply if treatment before release is attempted to reduce BOD below 30 mg/l. The wastewater treatment plant normally caters to a BOD level of about 30 mg/l for the effluent, thus calling for dilution of released wastewater through 'environmental flow releases (EFRs)' by a factor of ten if possible at the point of release. Such EFRs are very expensive particularly when there is unmet needs for other critical uses. Ideally, the treated effluent or wastewater with BOD higher than 30 mg/l and up to 100 mg/l could be led directly for irrigating agricultural farms, rather than releasing it in river systems and then requiring release of diluting doses of freshwater as EFR. Such action saves the precious diluting dose. Obviously, an economic method for treating wastewater below 30 mg/l is necessary, if it has no agriculture demand. Meanwhile, the solution lies in direct exchange of treated wastewater with desired inflow for domestic purposes. Case studies of degradation mainly due to industries, and regeneration of freshwaters of rivers Thames, Rhine, and Danube from Europe are often quoted. None of these rivers has an earmarked EFR. Also, these rivers are fed mostly by snowmelt and are not comparable with many of the monsoon-fed rivers. In ultimate analysis, the issue of EFR depends upon economic trade-offs rather than eco-regeneration.

The quality of river water in India for example is classified for different purposes: Class A with disinfection but without treatment. Class B for outdoor bathing. Class C requiring conventional treatment with disinfection. Class D is considered appropriate for propagation of wild life, Class E for irrigation, industrial cooling, and for waste disposal. Recognizing the deteriorating quality of river waters and the river Ganga in particular, the Ganga Action Plan (GAP) Phase I was launched in the country in June 1986. It mainly addressed interception, diversion, treatment of domestic-industrial wastewater from entering Ganga. Phase II was launched in Feb. 1991. The National River Action Plan (NRAP) was launched by India on similar lines for eleven other rivers. The Central Pollution Control Board (CPCB) identified nineteen grossly affected and fourteen less polluted stretches for treatment. The Sabarmati river, representing the case study in this chapter, is one of these rivers.

A river in a pristine condition carries a flow volume, which varies in time and space in response to the rainfall-run-off relationship for the sub-basins while supporting plant life and some biodiversity along the river basin. A hydrograph represents the flow pattern. Due to human interventions, the hydrograph representing say an average year flow, undergoes a transformation, which could affect the needs of the riparian society to the extent water is used consumptively in the upstream. A water resource development planner has to take care of such downstream needs adequately by means of direct supplies from abstractions made in the upstream. Most of these riparian needs like drinking and bathing requirements of people living near the rivers and for their cattle and livestock, are small and dispersed. Gradually, the organized water supply schemes are replacing even these needs. The sooner it is done better will it be; otherwise these have to be provided through maintenance of minimum flows, which again are very expensive. The adoption of the concept of satisfying a minimum flow need (MFN) along the river length is often insisted by some. The quantum to be provided however is neither assessed nor prescribed. The rivers fed by rainfall spanning say only from twenty to fifty days in a year, do not carry any flow for most of the non-rainy season, excepting perhaps some stretches in delta regions carrying the regenerated flow. For such rivers, there is no reason for release of MFN in dry season all along the river length as it is likely to be lost due to evaporation and/or seepage. The downstream riparian needs, therefore, ought to be taken care of through direct organized supplies rather than expensive releases in river system, entailing loss which often goes up to 60 per cent of the releases, depending upon individual cases.

The MFN sometimes identified at about a certain proportion of the flow in pristine conditions cannot be justified. The available flow comprises certain regenerated flow contribution from that used for irrigation. Similarly, wastewater released by domestic and industrial sector is to be counted towards the minimum available flow. The quality of the regenerated flow is of doubtful nature unless it is treated before release. The EFR and such regenerated flow have to be more than the MFN. Of late, the EFR and MFN are lumped together and called as committed flow (CF) for riparian human and ecosystems. Unfortunately, the concept is still in an evolutionary stage but is often held sacrosanct by some, causing avoidable loss of precious water needed for more critical needs. Clear assessments for different agro-climatic regions, including seasonal river systems are undoubtedly called for.

SUSTAINABILITY

Ancient habitats naturally rose around sources of water. The situation changed after the industrial revolution, causing a considerable unevenness in their distribution. Human intervention redistributed the naturally available water in the basin to suit the needs of such unevenly distributed settlements. Such a process has to be based on the cornerstones for socio-economic development viz. efficiency, equity, economy, and efficacy. It entails sharing of available water between surplus and deficit areas. Essentially, while sharing water and benefits due to development, different regions in a basin must, as well, share disadvantages like loss of livelihood, resources of land, homesteads, and forests falling to the lot of some habitants of the basin by integrating the dispossessed amongst the beneficiaries in one way or the other. Thus, IWRDM should result in a win-win and financially sustainable proposition for the basin stakeholders. Financial sustainability calls for adoption of principles of 'users pay' and 'polluters treat before release' or else 'pay', with requisite modifications to suit economic status of a country. Appropriate pricing mechanisms for services offered constitute a major determining factor for achieving such sustainability.

Integration also envisages site-specific adoption of sizes of infrastructure from mega to micro scale, from traditional to modern, their planning, design, construction, safety, operation and maintenance (O&M), modernization, rehabilitation, replacement; from considerations of availability, supply, and demands; augmentation, conservation, recycling, treatment, run-off to sinks and oceans. IWRDM has to deal with science and technology (S&T) and engineering inputs; social, economic, environmental effects—both positive and adverse; administrative, financial, legal, institutional mechanisms and aesthetic, cultural ethos. From ancient times, the state has undertaken water services to the community as its sacred and often a religious duty. Down the line, with urbanization and later with industrialization, the subject became complex needing more organized and centralized effort, making it inseparable from the state. Community involvement remained at local level. With local self-governments gradually coming to the fore, community participation became important. The clock has turned a full circle and now community participation in all socio-economic sectors of human development is considered essential. But, little work has been done on developing instruments for operationalization of these concepts. A movement of NGOs is engulfing the development scene. It unfortunately remains restricted to actions at the local level or anti-establishment movements. At basin level, some developed countries have launched authorities under community control but little evidence is yet available about success stories of effective community participation for total IWRDM from the developing world.

While achieving IWRDM one has to keep in mind optimization of fruition, cost-effective investments, adoption of correct pricing norms consistent with the ability of the beneficiaries and returns to society for ploughing them back in to the economy to enable undertaking further programmes for socio-economic development. The integration envisages sustainability of development and not sustained underdevelopment.

Because of uncertainties in climate, it has to be resilient and flexible enough to cater to extremes faced during natural disasters with inescapable variations while ensuring building of the society's capacity to bounce back once the adverse situation passes.

GLOBAL PERCEPTIONS SUPPORTING IWRDM

After the second WWF was held in the year 2000, the ICID which is one of the leading professional bodies, and was one of the main partners in developing the 'water for food' vision, adopted its 'strategy' for implementing actions emanating from the Vision. The strategy considers that ideally, supply side of IWRDM must be planned basin-wide. Infrastructure development should follow such plans, some purists say. But the world over, there is possibly no river basin where the concept has been practised in totality. Improvize as you build, and improve as you go along, has been more or less a rule in water resource engineering, which deals with natural resources with all their vagaries for the benefit of humankind and possesses a range of complexities particularly in the developing world. Undoubtedly, initial development takes place on the basis of felt needs. The basin concept, therefore, is mostly tested post facto or midway, to reduce a drift away from the ideal during practice and/ or for identification of necessary mid-course corrections. Evolution of basin plans more or less takes a period of time, making them useful for filling gaps in development schemes adopted till then. Fully successful basin-level agencies achieving IWRDM and to serve as models to suit the variety of situations are almost not available. All the same, the concept has found wide acceptance round the world and is being tested on ground with reports indicating varying degrees of success.

Flowing from the strategy, ICID has launched a Country Policy Support Programme (CPSP) to incorporate the IWRDM concept. For supporting evolution and adoption of appropriate water policies in developing countries, the CPSP is being conducted in five representative countries. Initially it has been started in China and India, the two most populous countries presently accounting for 38 per cent of the world population and 41 per cent of the world's irrigated area, where a good deal of basic data are incidentally available. Consultations will follow in Egypt, Mexico, and Pakistan. Sub-Saharan Africa is currently experiencing food insecurity, but because of lack of adequate data, countries in that region will be taken up for studies later.

While population grows, the irrigated area in these countries will also grow. In the ultimate stage, as the growth trends suggest, the two countries will probably represent 42 per cent of population and 45 per cent of the world's irrigated area. The two countries are progressing fast in several directions. Although presently agrarian, the societies are fast getting urbanized and industrialized. Both are wedded to the principle of self-sufficiency in food, fibre, and energy production. They have several in-country shared river basins along with some international basins, yet have significant unutilized water resource potential and need significant water resource development and require inter-basin transfer of waters. They both have ancient water-centred civilizations and a national policy in favour of IWRDM.

India has a National Water Policy first enunciated in 1987 and recently revised in 2002. An action plan for its implementation is on the anvil. China similarly has national laws on water, flood control, and water and soil conservation. Both have a well-articulated and largely successful irrigation policy. The CPSP therefore plans to utilize their existing knowledge base. The knowledge base will be enriched, based on assessments in two river basins each, one a water-deficit, the other relatively surplus for enabling a synoptic view of IWRDM in all the basins of the countries. The assessments will enrich the 'river basin information systems' to focus stakeholders' attention on the critical issues. Thereafter, consultations with stakeholders at basin level will be held on the basis of such scientifically developed knowledge base. They will be followed by national consultations for identifying and highlighting inputs for the country policies. Discussions are planned with the concerned national and provincial governments and funding agencies like World Bank, Asian Development Bank, Japan Bank for International Cooperation and others to facilitate development of appropriate policy interventions for the future. Incidentally, both these major players are presently in the process of evolving their own strategies for the water resource sector. In India, the two basins selected are Sabarmati on the west coast and Brahmani on the east coast. The present paper is based on the ongoing assessment for the CPSP in the Sabarmati basin.

The effort in these two countries will be followed up in three other countries namely: Egypt, Mexico, and Pakistan. Egypt is a unique case for study as it gets most of its waters from abroad through one river basin. But it has successfully finetuned its water policy and is making strenuous efforts to meet the needs of the growing population through integration of different facets of IWRDM. Pakistan similarly is a country dependent on most of its waters from the Indus basin. A significant share of its waters comes to it from India because of the 'Indus Water Treaty'. It is also engaged in making IWRDM a success. Mexico is located in the western hemisphere, has several river basins, and has evolved an integrated approach based on irrigation management transfer to farmers successfully. The CPSP is thus expected to serve as a representative package of water-related policies for adoption by the developing world.

India's Water Resources and the Sabarmati Basin

OVERVIEW

India has 2 per cent of the world's geographical area, 16 per cent of the population, but 4 per cent of freshwater resources. Its geographical area is 329 Mha of which 47 per cent is cultivated, 23 per cent is forested, and another 23 per cent comprises wasteland. It has twelve major river basins, which contribute about 85 per cent of utilizable water resources with the area of each being more than 20,000 km^2. The rest of the numerous small and large basins contribute the remaining 15 per cent. Forty-six medium basins have areas between 2000 and 20,000 km^2 and fifty-five minor basins have an area less than 2000 km^2. (Figure 3.1). All in all, the average annual rainfall emanating largely from the south-west monsoon is 117 cm, which is higher than the world average of 110 cm. The Himalayas get some additional rainfall in winter due to westerlies, north-east and south India due to the north-east monsoon.

Depressions that develop in the Bay of Bengal and the Arabian Sea play a major role in the precipitation pattern in South Asia and often cause floods and cyclones. The westerlies bring snowfall in the Himalayas. The snowmelt due to summer causes a second flood in the Himalayan rivers, making them perennial. Rainy days in the country are about 100 in a year. At places in the north-east, however, it rains for 300 days in a year, but at others in the north-west only for about ten days. The minimum to maximum proportion of incidence of annual rainfall in India is 1:100, the average being about 10. But the variability in time and space is large, the intensity is high, and a large portion is precipitated in about 100 hours in a year. Himalayas in the north, the hills along the east and west coastlines, and the ranges at the edge of the peninsula form the main orographic influences for the variability. They also determine the river basin boundaries.

About 35 per cent of the geographic area of the country is drought prone, whereas 20 per cent is flood prone, while 47 per cent is arable. Often, the same area is affected by both the extremes within one monsoon season causing water management problems. The river basins are covered

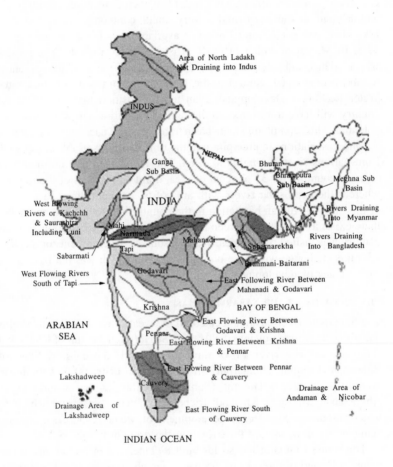

Figure 3.1: River basins of India

by twenty-four groups with physiographically identical behaviour (Figure 3.1). They annually receive about 4000 billion cubic metres (BCM) of water and drain the unconsumed water from the continent into the oceans. The country gets another 400 BCM of water from across its borders with Nepal and Bhutan in the north and a very small component is exported to Myanmar. All in all, annual average availability is however assessed at 1900 BCM, out of which 1100 BCM is considered useable. The present water balance indicates that about 58 per cent of the available potential has been developed. Recently, the National Commission on Integrated Water Resources Development Plan took a position that, by 2025, the country will have to develop all the remaining intra-basin water resource potential, while completing inter-basin transfer where necessary, afterwards.

Due to variations in precipitation, the run-off and its usable component from various rivers is highly skewed. The Himalayan and north-eastern rivers accounting for 33 per cent of area, drain 62 per cent of the run-off, whereas the peninsular rivers with area of 67 per cent drain 38 per cent of the run-off. Human habitation in the country is also skewed and hence there exists a mismatch between needs and availability. Not long ago, the country was visited by famines. Water resources development for meeting with this challenge was therefore the first task, which was realized in the early stages of development.

PHYSIOGRAPHY OF THE SABARMATI BASIN

The Sabarmati river basin is located on the west coast of India and drains the arid/semi-arid region into the Arabian Sea through the gulf of Khambhat (Figure 3.2). The river flows through the arid to semi-arid State of Rajasthan (Udaipur, Dungarpur, Sirohi, and Pali districts) and semi-arid drought-prone north Gujarat (Sabrkantha, Banskantha, Gandhinagar, Ahmedabad, Mehsana, Kheda districts). The latter is better endowed with water resources. Significant development of water resources has been achieved by now, mainly for irrigating the drought-prone area.

The state of Gujarat has so far built eighteen major and eighty-seven medium irrigation projects in the basin, which store about 1471 million cubic metres (MCM) and has plans to build another eight major and forty-eight medium projects, which will store an additional 150 MCM of water. A 'major' project is one that irrigates an area more than 20,000 ha and a medium one irrigates between 2000 and 20,000 ha of arable land. The state's harnessable surface water resources are estimated at 27.4 BCM. In addition, the interstate Narmada development will contribute 11 BCM with some more from the Rajasthan-Sabarmati link of the proposed inter-basin

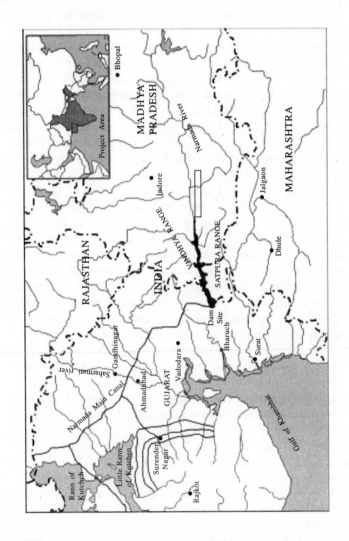

Figure 3.2: The Sabarmati river basin

linking project. The ultimate surface water irrigation potential of Gujarat is 2.2 million hectares (Mha) plus groundwater potential of about 2.5 Mha. Out of this, potential of 1.6 Mha has been created. The remaining is likely to be completed by 2007. In addition, Narmada will enable irrigation of about 1.8 Mha.

The Sabarmati river basin is a medium-size basin, with a drainage area of 21,674 km^2 (2.2 Mha) but carries the lowest run-off per unit area among the twenty-four groups. A part of this area of about 4124 km^2 lies in the relatively very arid and sparsely populated region of upper riparian state of Rajasthan. The basin is elongated in shape in north-north-east to south-south-west direction with a length of 371 km, taking its origin at an elevation of about 762 m above mean sea level (msl) at North latitude 20 deg-40 min and East longitude 73 deg-20 min. It has a maximum width of 105 km. Sabarmati has three major sub-basins, viz. Dharoi, Hathmati and the Watrak (accounting for 11 per cent, 46 per cent, and 43 per cent of area respectively). Land use comprises 9 per cent forests, 18 per cent waste land, and 73 per cent arable area. The river rises in the Aravalli ranges of north-central India comprising igneous and metamorphic formations of pre-Cambrian age. The main limb of the river runs through hilly formations for about 100 km accounting for nearly 25 per cent of the basin area, entering alluvial tract thereafter accounting for the remaining 75 per cent area, which includes the river delta filled up to a depth of about 2600 m. The river flows with a bed gradient initially of about 2 m/km, then with 1 m/km reducing to 0.5 m/km in the delta region.

The river basins of Banas, Khara, Saraswati, Khari, and Rupen lie on its west, separated by a low ridge, draining into the desert (Little Rann of Kachchh) and disappearing. The sub-basin of Chambal (Yamuna-Ganga) is in the east and the Mahi in the south-east. The Little Kachchh desert has a narrow neck which enables it to discharge flood flows, may be once in five years into the Gulf of Kachchh. The river Bhogavo with an independent drainage area from the western Saurashtra peninsula of Gujarat, meets the estuary of the river from the west close to the mouth and hence is not considered as a part of the basin in this case study, although technically it is. The main Sabarmati river course hugs its western boundary after leaving the Rajasthan state. The Rajasthan rivers: Wakal with a length of 38 km and Sei with 45 km join the main river from left and right at 51 and 67 km from its origin. In Gujarat, three tributaries viz. Harnav (972 km^2 drainage area, 75 km length), Hathmati (1523 km^2 drainage, 105 km length), and Watrak (8638 km^2 drainage, 248 km length) all join it from the east at 106, 170, and 303 km from the origin.

The Sabarmati basin comprises ten smaller sub-basins; two in Dharoi, three in Hathmati, three in Watrak, and two in lower Sabarmati. The Sabarmati basin's annual average rainfall is 753 mm, the Watrak sub-basin being the wettest with 812 mm rainfall and 40–60 per cent coefficient of variation. The natural rate of evapo-transpiration at about 12,000 MCM is high, the rest results into average run-off of only about 24 per cent, equal to about 2300 MCM. The 75 per cent dependable run-off is about 1539 MCM, indicating the aridity of the region. The river system carries flow during the latter part of monsoon, leaving the river regime dry through most of the year. The climate ranges from semi-arid to arid with temperatures in summer rising to 43°C and in winter falling to 3°C. Dust storms occur during summer with thundery development once in a while. Depressions in the Arabian Sea sometimes develop into cyclones which affect the southern delta of the basin.

The natural recharge in rocky area is small, whereas in alluvium it is high due to sandy-loamy nature of soils. Surface water resource of the basin is estimated at 1694 MCM on the basis of flow observations at Nabhoi, a place about 10 km upstream of the mouth of the river in Arabian Sea. The average overall natural recharge to groundwater is estimated at 2570 MCM. Utilizable recharge is estimated at 2096 MCM. Out of the three sub-basins, two in the upstream, face mining of groundwater from the static resource, which is estimated to be quite large. In the third sub-basin, bordering the gulf of Khambhat, however, there is a build up of groundwater mound due to surplus availability from Watrak sub-basin and recharge from Mahi irrigation. As a result, there is no salinity intrusion in this area as compared to the other coastal tracts of the state of Gujarat.

The strategy for the use from the three sub-basins has to change to reduce mining from the upstream sub-basins and use more groundwater from the downstream sub-basin, where it is surplus. Legislation on restricting use from groundwater has been introduced in a bill pending approval of the legislative assembly of the state for quite some time. But such envisaged restrictions have by and large failed worldwide. Real solution lies in providing requisite quantity from surface water, even by resorting to inter-basin transfers or else to make power supply expensive, so that surface water is preferred. The average water availability per capita as per the census of 1991 is about 360 m³, which is the lowest amongst India's basins. With growth in population, it is expected to worsen. Notwithstanding the natural deficit in availability, the case study presents how it can be ameliorated in future.

RURAL-URBAN HABITATIONS, AGRICULTURE AND MIGRATION

Human population of the basin at present is about 11.44 million (M), out of which urban population is 5.7 M and rural about 5.4 M, with respective growth rates of about 2.8 per cent and 1.1 per cent per year. The population is likely to be about 19.47 million in 2025 which could grow and may stabilize at say 24 million by AD 2060. In 2025, the urban-rural composition might be 12.27 million and 7.2 million respectively. Urban habitats are concentrated in southern plains near Ahmedabad and the state capital Gandhinagar, accounting for 35 per cent of the basin population. Other twenty-five urban pockets comprise the cities of Danta, Dehgam, Daskroi, Himatnagar, Idar, Kapadwanj, Khambhat, Khedbrahma, Malpur, Meghraj, Mehmadabad, Modasa, Nadiad, Petlad, and Prantij. The cities of Anand, Borsad, Dholka, Karol, Kheralu, Thasra, and Vijapur partially lie in the SM basin.

Industries are similarly non-uniformly concentrated. The growth of population is more in urban areas because agriculture cannot support it, and migration from rural to urban areas continues. Agriculture and cattle rearing for milk and milk products are the main occupations in rural area. Cattle rearing has become an organized sector, with dairy development prospering during the last forty years. Average agricultural land holding per farmer is 0.5 ha. But with each generation, fragmentation continues which is to some extent neutralized by migration. Sizeable proportion of rural population remains landless and subsists on employment offered by larger farmers. Like elsewhere in India, land reforms coupled with alternative employment generation in rural areas holds the key to economic revival.

HEADWATERS TO DELTA

The hilly area is largely denuded and covered by sporadic sand dunes, scrub, and a little sparse forest. The land with medium slope has shallow soil mantle, is mostly rain fed and is habitated by farmers practising subsistence agriculture. Where agriculture does not produce enough dried biomass, firewood needs are satisfied by cutting down of bushes and the degraded forests, thus accelerating denudation. In the event of the failure of the monsoon, migration of landowners resumes along with the landless in search of the employment. In rain-fed areas, the vagaries of the monsoon coupled with population growth and poverty constitute the largest cause for degradation of environment. Lower down, the plains are fertile but rain-fed agriculture has low productivity. Where irrigation water is provided, more productive agriculture becomes possible. The Sabarmati delta head lies in the urban-industrial centres around Ahmedabad and Gandhinagar. The rapid urban spread in this area has generally reduced the land availability for agriculture and fertile land in particular, putting constraints on both. Further lower down, the Sabarmati delta on the left bank merges with the plains of Mahi river on the river bank, which have been receiving irrigation waters for the last four decades. Area under crops is 1.56 Mha which forms 73 per cent of the basin area, out of which 0.15 Mha is irrigated by surface and 0.52 Mha by groundwater. Electricity from power grid is used to run the groundwater irrigation pumps for about 70 per cent of the area; the rest use diesel pumps.

DELTA, GULF, AND TERMINAL RESERVOIR

Further lower down, the plains merge into coastal saline and marshy lands on the periphery of the Gulf of Khambhat. Tidal zone starts at Motiboru village of Dholka *taluka*. On the right bank of Sabarmati, the river Bhogavo has a common delta and coastal saline lands. On the left bank, the Mahi and Sabarmati plains merge into common coastal saline lands. There is no distinguishing ridge between the two, but roughly 50 per cent can be attributed to each of the two rivers viz. Mahi and Sabarmati, respectively. The spread of irrigation has stopped there because of flat slopes leading to congestion of natural drainage. The agricultural drainage also is affected due to this constraint. The Gujarat state government had a separate Khar (Saline)-land Development Board, which attended to development of the coastal saline land. So far, very little proportion of such land has been developed, reclaimed, and distributed on lease basis for agriculture, basically for want of freshwater. The technique used comprises protecting the coastal front with embankment, providing sluices to prevent saline water advance in tides and to enable draining out the inland floodwaters.

Serious efforts to go in for reclamation have not been possible. Shortage of freshwater for leaching salts has possibly been the main reason. With the multipurpose Kalpsar project envisaging a terminal freshwater reservoir across the gulf to include residual waters of rivers on both sides of the gulf, reclamation efforts could get a boost. Also, with operationalization of the Narmada canal, more freshwater would be available for leaching of salts. The plans for Kalpsar reservoir include, besides Sabarmati, the mighty Narmada and Mahi rivers flowing into the gulf from the east. Pre-feasibility studies for the project have been recently completed and feasibility studies initiated. The project aims at reclamation of coastal saline land out of which about 20 per cent might be from the Sabarmati basin, besides providing irrigation facilities through a garland canal around the gulf. A tidal power station to utilize the country's highest 11 m tidal range in the gulf, is envisaged to generate hydropower.

IWRM in Sabarmati Basin

SURFACE WATER IRRIGATION

The first main diversion structure of the basin was constructed in the Dharoi sub-basin in the 1920s to enable protective irrigation on the river bank. Growth of surface irrigation since then progressed slowly, but picked up from the 1950s. The first reservoir in the basin, Meshwo, was completed in the 1950s storing a gross volume of 75 MCM and live storage of 72 MCM, enabling for the first time irrigation over two seasons. So far, 1471 MCM of gross storage has been built in the basin. Besides, future addition of about 150 MCM has been planned. The present live storage is about 1336 MCM. Storage for unit availability in Dharoi sub-basin is 129 per cent, indicating a carryover to capture all the water that was available. It is however only 38 per cent and 15 per cent in Hathmati and Watrak sub-basins indicating scope for future development. For the entire basin, the ratio is 42 per cent indicating undeveloped potential of 58 per cent. The cropping intensity in diversion-based irrigation is 0.75, whereas it is of the order of 1.2 for storage-based schemes. Main canals, branches, distributories cover about 1000 km length. The minors and field channels cover may be 7000 km. Surface irrigation area at present is 0.15 Mha, the groundwater coverage being 0.52 Mha. Present annual use of surface water for irrigation is about 2800 MCM. It is expected to rise to 3243 MCM in 2025. Irrigation map of the basin is indicated in Figure 3.3.

The main surface canal systems lying in the three sub-basins are as follows: i) Dharoi sub-basin—Right Bank Main Canal, Left Bank Main Canal Harnav I, Harnav II, Sabarmati Reservoir Project; ii) Hathmati sub-basin—Guhai, Hathmati, Fatewadi, Kharicut; iii) Watrak sub-basin—Meshwo, Mazam, Watrak, Meshwo canal, Waidy. An area of about 15,000 ha is irrigated by deployment of treated sewage wastewaters in Fatewadi canal system. There is a significant demand for supply of such treated wastewater, which augurs well for augmenting treatment capacity.

It will reduce the ecosystem degradation at the same time. Main food crops grown in the irrigated area in the south are paddy, wheat, maize, pulses, and groundnut. Cash crops grown are tobacco and vegetables and fruits. In the north, the food crops are bajra, jowar, oilseeds, and cash crops are cotton, cummins, and certain spices.

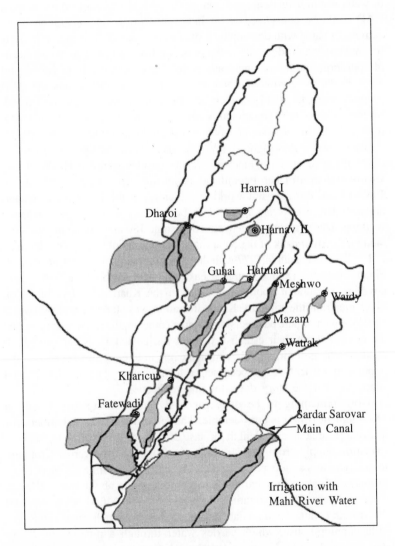

Figure 3.3: Irrigation map of Sabarmati basin

Command Area Development (CAD) activities were introduced in the basin from the 1970s, keeping in view the predominance (about 90 per cent) of water use in irrigation, much before the IWRM was adopted world over. CAD could be considered as an indigenous effort for integration of all activities related with development of command areas of irrigation projects for maximizing productivity and production from irrigated areas, by means of appropriate structural and non-structural interventions. It, of course, did not consider integration of water use in different sectors. Under the CAD programme so far, field channels and field drains have been constructed, whereas *Warabandhi* (rotational water supply) has been adopted in a very little area. So far no area has gone out of production due to waterlogging.

Water Users Associations (WUAs) have been set-up by farmers to cover irrigated areas in Fatewadi canals besides some in Hathmati and Dharoi system. In Dharoi command, about fourteen WUAs were set-up in 1983–84 but they are not working presently. In the Fatewadi command however ten WUAs are reported to be working and two have stopped working. This irrigation system covering some 180 tanks lying in low-lying land on the right bank of the Sabarmati is 100 years old. It was modernized in 1967–68 and again in 1977 with the commissioning of the Wasna barrage. Finally, with the Dharoi system it has become a reliable irrigation system covering an area of about 36,000 ha. In the Kharicut system based on diversion from the tributary Khari, irrigation was started about 300 years ago. When an organized diversion scheme came up, the farmers approached the court and got their established rights designated as 'Kalambandhi'. It resulted into an informal WUA, which came in to existence because of the popular movement. It has been active for all these years and has proved successful but the story does not find replication.

Work remaining to be done for CAD and Participatory Irrigation Management (PIM) is considerable. The left-bank area of Sabarmati, for a sizeable area over a length of last 50 km of river length, received irrigation supply from Mahi river basin from 1970 onwards. Cropping intensity in this area is presently 1.3. Presently, the Mahi right bank canal provides irrigation in Sabarmati basin amounting to about 1800 MCM of water per year. This quantum is besides about 100 MCM being supplied for drinking water to Ahmedabad through Shedhi branch canal and Raska village pump-house, which carries water through a pipeline of 2.1 m diameter and about 32 km in length to Kotarpur water treatment plant. About 60 MCM of water is exported to adjoining area across the basin boundary through the right-bank Dharoi canal. Another 35 MCM is exported out of the basin area from Wasna barrage on the right bank.

Recently, the Narmada main canal construction has been completed well past the Sabarmati basin. The canal crosses Sabarmati river upstream of Ahmedabad after covering the whole of the Sabarmati basin irrigated by the Mahi canal. The construction of branches and the distribution network of the Narmada canal will be completed by 2005, when regular irrigation is expected to start in Sabarmati command. About 140 MCM of water per month (about 50 m^3 for 100 days) has been released from the Narmada to the Sabarmati river since August 2002 to take care of the drinking water needs of affected villages due to the century's worst drought in the downstream and of Ahmedabad city proper.

Presently, about 365 MCM/year of Narmada water is being regularly used in Shedhi and Raska command areas of the Sabarmati basin. The utilization is expected to be about 2200 MCM in 2025, comprising additional 688 MCM in Hathmati and 1351 MCM in Watrak commands. The figure of 2200 MCM includes regular supplies through Narmada main canal round the year and use of monsoon surpluses which will be available for a fairly long term. This is in addition to about 100 MCM for drinking water. Present plans are to release Narmada waters for irrigation in the Mahi/Sabarmati command to take care of the supply of about 1000 MCM in a year, through the existing distribution system. In the short run, it is expected that irrigation supplies will be augmented significantly through the waters of the Narmada. The long-term picture is yet to emerge. But, one thing is clear and that is import of Narmada waters is going to be an important component of irrigation water use of the Sabarmati. Synchronization of the use of Sabarmati, Mahi, Narmada waters for maximization of irrigation production from this area will now be possible. At the same time, better use of Watrak surplus has to be thought out. Along with Narmada waters, it could be used for northward transfer to the tune of over 1000 MCM of water per year, for irrigating needy rain-fed areas through monsoon spills, which are proposed to be diverted through the Narmada canals.

GROUNDWATER IRRIGATION

Dug wells were initially used for abstracting groundwater from unconfined aquifers in the basin for irrigation. Such aquifers still hold water during the monsoon and a part of the post-monsoon season. The development in the basin as a whole has risen to about 65 per cent. Below the unconfined aquifers lie the multilayered confined aquifers in alluvial plains, which are being mined. With surface water irrigation spread, recharge to groundwater started increasing, enabling more withdrawals. Increase in the availability of electric energy and diesel pumps revolutionized groundwater extraction. Presently, the basin has more than 40,000 tube wells for irrigation, besides those for drinking and industrial needs. But on the flip side, the extractions soon overtook the natural and surface irrigation recharge. Groundwater levels started falling, resulting in its mining. Cost of pumping mounted. Small farmers were put to a disadvantage. The situation however seems to be stabilizing. Present extraction from groundwater in the basin is about 2700 MCM/year. It is estimated to rise by another 1500 MCM/year by 2025, unless satisfied by surface water sources. The need for artificially recharging the depleted aquifers has been recognized. The possibility of recharging by means of a pair of shallow and deep tube wells has been proved. Besides, other measures include—underground check dams, percolation tanks, check dams, and single recharge tube wells. Little work of actual recharge has yet been carried out, mainly because it is expensive.

Analysis on the basis of sub-basin indicates that upstream of Wasna barrage, there is an adverse balance of groundwater use, whereas in the downstream a groundwater dome is being built due mainly to surplus availability from Watrak sub-basin. In the upstream, inflow/outflow across the basin boundary is not very well assessed and is under investigation. With the import of Mahi-Narmada waters, the situation is going to be more favourable in future. Conjunctive use will be increasingly put in practice hereafter in the downstream areas. The Narmada waters can undoubtedly play a major role in recharging the depleting aquifers in the upstream. The state is planning to take the monsoon surplus from Narmada canals to ponds along the unirrigated rain-fed areas and use them for artificial recharge. If the plans materialize in the future, the aquifers being mined might be recharged to sustain the groundwater irrigation in these upstream sub-basins.

Groundwater pumping for irrigation presently accounts for about 30 per cent of electric energy used in the state and yet energy prices are low. Power supply leaves much to be desired and is available in short spurts when the farmers are unwilling to use it. The demand always outstrips the supply. With commissioning of hydropower stations of Sardar Sarovar Project on the Narmada, the position is likely to ease out. It is expected that conjunctive use of surface and groundwaters will be possible with sustainability in the next decade.

RAIN-FED AGRICULTURE

The Sabarmati basin is drought prone; its aridity increasing towards the north. Its rural habitants are poor, belong to the backward tribes, and many are landless and often food insecure. They are affected practically every third year by a severe drought. During drought, even sown crops are lost. The landed and landless then migrate in search of employment and livelihood. Cattle owned by them graze on whatever little biomass that grows. Often in the absence of adequate supply of fodder, distress sale of cattle takes place, thus causing a loss of animal power for the next crop season. In absence of crop residue, people burn twigs and branches of trees thus causing ecological loss. Rain-fed agriculture remains a losing proposition for both the farmers and the ecosystem. Although irrigation is a must for supporting even subsistence agriculture, the needed water is simply not available within the basin. A part of the basin is likely to remain rain fed even in the ultimate stage of development of irrigation, with the imported water through Narmada canal system. The wasteland is also sizeable. There are plans for distribution of such land to the landless or for horticulture.

Historically, with the growth of population the cropped land has got fragmented with every passing generation, in spite of migration to urban and industrial sectors. With irrigation, more hands/mouths are supported on even the fragmented landholding. But in case of rain-fed land, encroachment on wasteland or forested land, takes place and attempt is made to bring it under plough without reclaiming the former or at a cost to the environment in the latter case. Land reforms are essential for the rain-fed areas, without which eco-degradation will continue without let up. Because the rainfall is unreliable, a farmer has to gamble, often losing out and causing loss of livelihood when crop fails. The agriculture remains at subsistence level increasing the pressure on wasteland or forest.

There are proposals to fill natural as well as other small ponds with surplus water from the Narmada in the rain-fed land. The massive-scale investment of funds being done for the watershed development programmes would possibly help some proportion of this rain-fed land. Finally, the interlinking of rivers being planned for the future might help bring another cultivated area under supplemental moisture regime from the Rajasthan–Sabarmati link canal. Thus the prospects are that in the ultimate scenario in the Sarbarmati basin, rain-fed area purely dependent on rains will reduce considerably.

Different models for development of rain-fed areas are being envisaged. One holding great promise, links developing biomass-based agriculture with participatory approach to make the dependent population get their livelihood while regenerating the degraded ecosystems. The participatory approach would help local determination of the stakeholders' resources, their priorities, while assessing their capabilities. Some others will be dependent on land reforms and yet some others on simply weaning away people from agriculture to some agro-based industry based on horticulture. Many claim that only a bottom-up approach will be successful, but experience tells that a mix of top-down and bottom-up approach is necessary to sustain progress. It has to be understood that rain-fed agriculture will continue to depend upon food-for-work programmes, improving livelihood, ameliorating poverty, increasing accessibility to food, and improving health of the concerned.

NEEDS: DRINKING, DOMESTIC, AND INDUSTRIES

Habitats with less than 10,000 people are classified as villages. Those having up to 25,000 are classified as towns, up to 500,000 as cities having municipalities, and those having more than 500,000 people become eligible for the status of big cities needing municipal corporations. Cities with population more than 100,000 people are classified as Class I cities. Classes II, III, IV go down up to 50,000 for cities, and 20,000 and 10,000 for towns. Urban habitats are expected to grow fast causing severe stress on water needs. Presently, about 40 per cent of urban population is served from surface water and surface-based groundwater systems. Standards for water supply vary from 40 litres per capita per day (lpcd) for villages going up to 220 lpcd for cities. Present annual basin-level domestic and industrial demand for water is about 510 MCM and 128 MCM, making up a total need for 638 MCM. Ahmedabad complex alone needs about 338 MCM, i.e. about 51 per cent of the quantum.

Presently, groundwater takes care of about 58 per cent of this requirement but is being mined. Groundwater production for twenty-five cities is about 245 MCM/year. About 30 per cent of population is not yet connected to organized water supply schemes. It is dependent on own tube wells abstracting about 43 MCM/year. In addition, rural areas draw about 85 MCM/year, of which 90 per cent is based on groundwater. Thus, in all, about 365 MCM/year is provided by groundwater. Presently, 730 million litres per day (MLD) of Dharoi water is allocated to Ahmedabad and Gandhinagar complex. For this purpose, 8.5 m^{3}s flow is released but about 5 m^{3}s flow is lost in transit by percolation or unauthorized lift by farmers for irrigation from river bed and evaporation till the water reaches city limits. To counter this loss, a direct pipeline from Dharoi reservoir to Ahmedabad was contemplated but the scheme is presently shelved. With the import of Narmada waters, the mining is likely to be offset.

In the year 2025, the total demand for the basin is likely to be around 1122 MCM, out of which Ahmedabad will need 645 MCM rising to about 57 per cent of the total. About 30 per cent will still be based on groundwater, which has serious problem of water quality due to inherent salinity (TDS), fluorides, and nitrates at places. (As per 1991 census, about 921 villages of the Sabarmati basin have been identified as no-source villages, making up 22 per cent of the state.) There are 2408 no-source villages in the Sabarmati basin, making up a total of 77 per cent of the total of the state. Sabarkantha accounts for 47 per cent of the 2408 villages in the Sabarmati basin and 84 per cent of the district's no-source villages. Many villages have hand pumps, while very few have piped water supply (w/s) schemes. About 250 villages are covered with regional water supply schemes. The villages have to provide for bovine needs as well. They could be quite high, say 80 lpcd for buffalows, 50 for cattle, etc.

Depletion of the water table suggests unsustainability of the source, which can be augmented only by increasing the surface supplies. This will take place due to watershed development and imported water from Narmada and/or Rajasthan-Sabarmati link. Where such supplementation cannot take place, the abstraction will have to be restricted or a crisis situation will have to be faced. Presently about 30 per cent of consumed water is considered to be in this sector. It is likely to grow to 45 per cent in the ultimate stage. Amongst the country's states, the rate of urbanization-industrialization is fastest in Gujarat, and aspirations of rural people match those of urban population thus closing the divide fast. Drinking water needs will therefore grow faster than the growth of facilities for evacuation and treatment of wastewater. All in all, the domestic and industrial needs are likely to grow to 1100 plus 200 MCM/year. Wastewater can be conveniently assumed at 85 per cent of this quantum, needing for urban sewerage, including latrines, a vast amount for infrastructure, of say Rs 19 billion. In addition, it is estimated that for rural areas, need for investment will be Rs 2 billion.

All treated wastewater should become an important component of irrigation water if not in rain-fed rural area, then in peri-urban area. It will have to be seen that no untreated wastewater is released in natural streams unless earmarked for lift and use in the downstream. At the same time, all the drinking water needs should be met with from upstream treated water without addition of wastewaters. Regional schemes for rural areas will have to be promoted rather than individual village schemes. All these will ensure the reduction in supply-demand mismatch with quality supplies.

The Sabarmati basin is characterized by heavy human influence in the south, whereas the influence is relatively small in the north. In, future, economic growth is likely to act as a trigger for the north as well. There are about 40,000 industries in the basin. About 12,000 of them are causing pollution. They are well spread in the medium-and large-scale sectors, which have investment of more than Rs 100 million and 5000 million respectively. Small industries (investment below Rs 10 million) are mostly non-polluting. Textile industry dominates 75 per cent of large-scale and 50 per cent of medium-scale units. Next major component comprises chemical industry, which accounts for 11 per cent and 17 per cent of large and medium industries respectively. Wastewater emanating from the polluting industries is of real concern. Industrial units have to be encouraged to treat their effluent and then release it into the river. With the import of water from the Narmada, good dilution is now available. The need for EFRs therefore is not expected to arise in future. Ahmedabad treats about 85 per cent of its domestic wastewater. With supply needs tripling by 2025, heavy load will come on treatment and recycling of wastewaters. The conditions in other urban habitats and likely quantity of industrial effluent that will be generated are presently not clear. Probably, little effluent may be treated there. About twenty industrial estates of Gujarat Industrial Development Corporation (GIDC) are provided with water supply: seven in Ahmedabad, one each in Banaskantha and Gandhinagar, six in Kheda, two in Mehsana, and three in Sabarkantha catering to 7.5 MCM of water supply yearly. Treatment of their effluent is to be looked after by GIDC.

Water quality in the reservoirs is very good as evidenced by monitoring at five major reservoirs. The quality of Sabarmati's water is monitored at eight places along the river downstream of Dharoi. As the river enters the Ahmedabad city area, the water has BOD at 1.4 mg/l and dissolved oxygen at 8.3 mg/l (A class) with a little nitrogen load due to upstream agricultural chemicals. Natural dilution of this load does take place. Ahmedabad generates 740 million litres per day of wastewater, of which about 85 per cent is at least treated to primary level enabling its use in agriculture. At Vautha, at the end of the polluting area, the values are respectively 34 mg/l and 2.7 mg/l (E class). The Ahmedabad river stretch is one amongst the eleven included in the National River Action Plan (NRAP), mainly due to the municipal and industrial effluents, washing and dyeing of textiles, and the release of fly ash from thermal power stations. The pollutants persist till Watrak meets the river, where dilution is substantial.

The NRAP treatment comprises interception and diversion of sewers to treatment plants, desilting of sewers and storm drains, rehabilitating old and installing new sewage pumping stations, adding new treatment plants, and low-cost sanitation for slum dwellers. The state environmental action plan has also been mounted with assistance of the World Bank. With the inflow of Narmada waters, the work of NRAP will undoubtedly be eased. Secondly, the dependence on Dharoi water to the tune of 268 MCM/year with average supply of 150 lpcd in five zones of Ahmedabad is likely to be a thing of the past. In fact, with the arrival of Narmada waters, gradual phasing out of Dharoi supplies will be on cards sooner than later. Naturally, it will be useable in irrigation command which has been deprived of the supply because of priority bestowed on drinking water needs.

FLOODS

The flood-prone areas in the Sabarmati basin are restricted in the vicinity of Ahmedabad. Dholka taluka in Ahmedabad and Borsad taluka in Anand districts account for more than 25 per cent of the chronically flood-affected areas. In addition, Daskroi taluka in Ahmedabad district besides Matar, Mehmadabad, Nadiad, and Thasra talukas in Anand district are affected by recurrent floods to the extent of 10 per cent of the area. Floods and the areas affected are decreasing in intensity and frequency with the construction of upstream reservoirs. With proposed reservoirs coming up, the area will diminish further. On the other hand, with rapid urbanization and industrialization, more and more low-lying areas get filled up, get paved, and cause rapid run-off and more inundation. On the balance, effects of flood are likely to reduce in the basin. It is interesting to note that there is a proposal to jacket the Sabarmati river through Ahmedabad, for developing the river front. The proposal will take care of floods of high frequency and reduce inundation. At the same time, the valley storage will be affected, causing possible elevation of high flood levels.

FORESTS, BIOMASS, AND MANGROVES

The states of Gujarat and Rajasthan have about 9 per cent and 7 per cent area respectively under forests against the national average of 19 per cent, but both states indicate a trend of increasing coverage. The forest classes by definition include dense, open, scrub, mangrove, and non-forest areas. In Gujarat, pastures/grazing lands plus tree crops/groves, and wasteland respectively account for 5 per cent and 30 per cent. In Rajasthan, the figures are 5 per cent and 28 per cent respectively. These are likely to be converted at least partially into agricultural farms for settling weaker sections of landless people and human habitats for cities, towns, and industries. The Sabarmati basin has reserved forests lying in the hilly tract, a major portion of which is degraded. Some area is under encroachment by farmers. Shifting cultivation is also taking a toll of forests. There are plans to settle the nomads outside forests on wastelands. Plans are afoot to cover wasteland with forests through social forestry. The plains have little by way of forests. In the delta, about 79 ha are under mangroves. But a major part of mangroves in the gulf region are outside the influence of the Sabarmati basin. Plans are afoot to stop encroachment of mangroves for fodder and firewood. It is expected that in the next twenty-five years, the mangroves will grow in their coverage by about 50 per cent. Riverine area is almost bereft of ecosystem and plant life worth the name. There are some grazing lands in the basin, but hardly anything grows there. Fodder largely comes from crop residue or is imported from outside the basin.

Studies for Sabarmati Basin

The Sabarmati basin happens to be one amongst the few most studied river basins in the country. As a result of these studies, a series of reports has been brought out, which are listed on the references list at the end of this chapter. Prominent amongst them are those by Government of Gujarat (CDO), Government of India (CWC), Tahal Consultants (Israel), Tata Energy Research Institute, Institute for Rural Management (Gujarat), Indo-French Cooperation, Ratan Tata Trust (INREMF), Gujarat Water Resources Development Corporation, National Environmental Engineering Research Institute, Central Pollution Control Board, and Gujarat Engineering Research Institute. The Central Designs Organisation of the state of Gujarat keeps all the relevant record to serve its needs, but the 'Gujarat Data Centre' is organized to serve as an archive. The recently concluded Indo-French programme has proposed setting up of a Sabarmati Integrated River Basin Information System (SIRBIS), which is under consideration of the government. The CWC with the help of Dutch assistance is setting up a decision support system through a river basin simulation model (RIBASIM). All these studies highlight the efforts made so far in meeting with the present needs, but underline current shortages, degradation of natural resources, and the rising needs and shortages due to rapid urbanization and industrialization. They are all, however, upbeat because of the successful initiation of import of freshwater from the Narmada basin through its main canal, and consequent reversal of the era of shortages and eco-degradation.

Present, Past, and Future

HINDSIGHT

Among the basins of India, the Sabarmati basin has by way of precipitation the lowest annual average availability as renewable dynamic resource. About 30 per cent seeps into the groundwater, 40 per cent is natural evapo-transpiration, and 30 per cent probably flows down the river system as run-off. Abstraction from groundwater is showing a net deficit, which seems to be getting mined annually in the upstream two sub-basins. Sub-basins of Dharoi and Hathmati are completely harnessed for surface waters, although some storages are oversized. The Watrak sub-basin has a surplus surface availability which flows down unutilized. It meets the Sabarmati, aiding dilution in fair weather of polluted waters released by domestic and industrial sectors. Incidentally it is interesting to see that although Sabarmati basin has the lowest annual average availability of water amongst basins of India, it is the most urbanized one with likely urban-rural proportion being 65:35. It helps prevent salinity ingress in the coastal tract unlike other deficit coastlines. It also aids sustenance of the mangroves in spite of the population pressure. It recharges groundwater, causing a surplus availability in the downstream part of the basin as far as groundwater is concerned.

Still, sizeable quantity of irrigation water is exported outside the basin in physically contiguous areas, whereas the import from Mahi and now Narmada has started to outweigh the overall deficit. Trans-boundary deployment of basin waters has been practised in this basin almost right from the beginning. At the same time, need for import for meeting with the deficit also was recognized. Unfortunately, it was delayed due to misdirected opposition to the Narmada project by some activists. The need has been underwritten by the final judicial verdict, enabling the recently achieved transfer. During 2002, for the first time starting from August, about 600 MCM of water has been imported through Narmada canal and released in the Sabarmati river. The release has been more in the nature of a trial, but starting from this year, the import is expected to grow and could be used for drinking water, for dilution, for flushing pollutants, and for irrigation. One immediate effect of the import has been improvement of water quality downstream, and adequacy of drinking and domestic water supply in spite of the drought.

In the past, availability of basin water has been inadequate. Groundwater has been mined. Freshwater needs of urban habitats and industry have not been fully met with. Irrigation water is regularly diverted from irrigation for drinking water needs during droughts. Surface water and groundwater quality has degraded because of lack of adequate treatment of wastewaters, overdrawals, and native salinity. The urbanization and industrialization is proceeding faster than before. Needs are growing. At the same time, irrigation has been suffering. Rain-fed agriculture remains at subsistence level and due to fragmentation of land and population pressure, seasonal and permanent migration continues. Agriculture and irrigation at the present levels are not adequate to create needed jobs. Deforestation continues due to the pressure of population and firewood needs. Mangroves in the delta region are small in size but need to be sustained.

PEEP INTO THE FUTURE

Like other basins, the development and management of water resources in the Sabarmati basin has progressed on needs basis in the face of funding constraint. The use of surface waters for irrigation was limited by the lay of the land and the sites for building reservoirs. That is how irrigation water is exported outside the basin. In fact, utilization within a basin is more notional because proximity of irrigable land is a more important consideration, while extending the facility and not the basin boundary itself. There is scope for undertaking some more reservoir schemes to add about 150 MCM of utilization for irrigation through development of surface water resources. Irrigation efficiency is low and needs to be improved substantially, by releasing waters for increasing cropping intensity and food production, improving livelihood for irrigators, and making agriculture an attractive proposition. Simultaneously, irrigation productivity is expected to rise by 20 per cent. If land reforms are carried out, larger holdings will aid the process. Formation of WUAs is expected to pick up, promoting water saving, with recovery of water charges.

Present annual food grain requirement in the basin is of the order of 2.3 million tonnes against production of only about one million tonnes. The requirement in 2025 is projected as 4.4 million tonnes and expected production will be only about 1.5 million tonnes, thus indicating a significant shortfall. Such basin-level assessments are being verified through models like PODIUM. Depending upon the capacity to maximize food production, the likely shortfall is to be reassessed. Obviously, within a country, local basin-level insufficiency is dealt with through national availability and is not a matter of concern.

The import of Narmada waters in addition to that from Mahi river basin is expected to remove shortcomings in a big way. In particular, augmentation of drinking water availability, artificial recharge of depleting groundwater basins, and flushing of the pollutants from the river system will be possible. Expansion of irrigation into rain-fed areas will become possible.

Meanwhile rainwater harvesting will continue with vigour in view of increased outlays and better organization of the watershed development programme. It will simultaneously take care of the eco-degradation caused due to population pressure.

Steps like community participation in decision making, planning, construction, O&M, rehabilitation, modernization, and replacement of facilities are included in the nation's water policy. Very little however has been achieved in the Sabarmati basin due to several reasons. The WUA movement is yet to take roots in the basin. Socio-economic-ecological considerations have been attended to in the basin as brought out in the foregoing sections. The on-going plans take into account the nuances of these aspects, but a lot remains to be achieved. Public hearings are now held for new projects in accordance with the laws. The provisions are strictly followed by the implementing agencies. This ensures that once taken up, a scheme goes on unhindered.

Conclusions

For the Sabarmati basin, IWRDM has been assessed in all its facets through different studies and summarized in this paper. To begin with, various nuances of definition of the acronym are looked into. A case is made out to say that IWRDM should not be constrained by basin boundaries as demonstrated by water use across the basin boundary from the beginning in this case. Still, basin is considered as a useful unit for assessments and planning use in needy areas of the administrative region. The freshwater availability from the basin itself is known to be inadequate to meet with various demands as reiterated by different studies; some commissioned and some done by independent agencies. To make up the deficit, import of water from across the basin boundary has been practised for quite some time.

The present supply-demand mismatch, for the first time, shows a distinct possibility of resolution. For the year 2025 too, there are signs noticed by various studies that the needs will be met with. These possible fruits of integration will have to be sustained by several legal, policy, planning, institutional, operational, and socio-economic changes, for which steps have been initiated. The revised irrigation act and the state water policy are on the anvil in context of the recently revised national water policy. With these changes, faster and more equitable socio-economic development can usher in sustained economic growth for human development while ensuring ecological sustainability. The case study clearly brings out that supply management remains a core component, in particular for deficit basins like Sabarmati and ignoring it is harmful.

Setting up of a river basin council has been suggested in the report of the Indo-French Cooperation Programme. According to them it should comprise—national administration and authorities, regional and local representatives, users and polluters, and the think tank. The report also suggests support of a programme design and implementation organization and a assemblage of working groups. The studies have identified several institutional and legal changes besides building political awareness. Framework water directive is advocated to enable these desired changes. The state government itself has extensively studied the need for amending the existing framework. A bill for amending the existing irrigation act into a comprehensive law has been pending adoption for quite some time. Hopefully, it will be adopted in the near future. The amended law provides for setting up of WUAs and creating enabling environment for them.

Irrigation water needs as per present estimates are 6204 MCM/year. For the year 2025, they are likely to rise to a level of 7200 MCM/year. It is estimated that there will be a deficit of the order of 4260 MCM of water/year, if no measures are taken to improve the situation. Not only the deficit can be reduced, but some surplus can be generated, if the following measures are taken in a phased manner. Now that supply management is taken care of, urgent attention is necessary for demand management.

1. Completion of the ongoing and contemplated storages.
2. Continuation of current import from Mahi basin.
3. Continuation of import from Narmada basin via Shedhi river and Raska weir.
4. Additional import from Narmada basin.
5. Lift of Narmada surplus monsoon waters by pipelines to Dharoi and Meshwo reservoirs.
6. Extension of Dharoi right bank canal. Operationalization of Dharoi left bank high level canal.
7. Kadana recharge canal running parallel to Narmada main canal in NE about 50 km away.
8. Diversion of Watrak surplus for use in rain-fed areas of north and north-west.
9. Treatment, recycling, and exchange of domestic and industrial waste-waters.
10. Release of Dharoi waters from municipal use back for irrigation.
11. Water use efficiency increase in designated stages of 5 per cent, 10 per cent say every 10 years.

12. Non-structural measures of appropriate pricing, volumetric supply, improvement of O&M, CAD, WUAs, water laws, R&D, public awareness, conflict management, and capacity building.
13. Improvement of rain-fed agriculture with water harvesting to take care of livelihood and soil, water and biomass conservation through public participation, while reversing migration.
14. Feasibility studies for the Kalpsar terminal reservoir in the Gulf of Khambhat.
15. Final import of water through the Rajsthan-Sabarmati link of the inter-basin transfer project being undertaken by the country.

Water resources development and management in Sabarmati basin has not followed any predetermined model like most of the other basins. The concept of IWRDM is evolving with actual progress on ground, to include solutions to the problems faced by those who were charged with the related actions by the society, those who have been benefited, those who have been left out of the process, and several independent observers, scholars, and academicians who shape things to come. This chapter attempts to identify various aspects of integration and examines the status, which is an on-going process. Whether IWRDM is implemented or not, or to what extent, if not why not, are not the questions whose answers will be found in this paper. What is clear is that the progress achieved in identifying problems, in defining solutions and working for achieving them is significant. The chapter brings out an optimistic perception that although local problems may continue, solutions to them are on the radar for the year 2025. The questions often raised about IWRM can be answered as follows in the context of Sabarmati basin.

1. Was the IWRM process adopted successful?: Partially.
2. Reasons for the success: supply management, import of inter-basin waters, insistence on IWRDM and not IWRM alone.
3. Constraints—uncontrolled growth of urbanization requiring withdrawal of irrigation waters, overdrawal of groundwater, lack of WUAs and participation by stakeholders, low level of pricing, and lack of volumetric measurement for irrigation.
4. Plan for removal of the constraints: CAD wing is attending to them.
5. Social, environmental impacts are taken care of. There is all-round better conditions.
6. Capacity building: No special effort was made, but the area has long traditions of co-operative movements and hence the stakeholders are aware of their rights and responsibilities.

References

Draft Presentation on DSS RIBASIM, New Delhi: Central Water Commission, 2002.

Effect of Pollutants in Sabarmati through Ahmedabad, (Draft Report), Vadodara: Gujarat Engineering Research Institute, 2002.

Government of Gujarat, *Integrated Plan of Sabarmati River Basin*, Gandhinagar: Central Design Organisation, 1996.

Gupta, S.K., *Sustainable WRD Plan for Ahmedabad, 2001 and Beyond*, Roundtable. Ahmedabad: 1997.

Integrated Sabarmati Basin Management Programme, Gandhinagar Indo-French Cooperation, 2001.

Kumar, Singh, *Integrated WRM in Sabarmati Basin, Issues and Options*, Anand: Ratan Tata Trust, (INREMF), 2001.

Parmar, B.J., and O.T. Gulati, *Sustainable Water Resources Development through IRBM on Sabarmati River Basin of India*, Gandhinagar: 1997.

Patel, M.S., B.M. Rao, N.J. Patel, *IRBPM, A Demonstrative Study on Sabarmati River Basin*, Gandhinagar: 2002.

Phadtare, P.N., *Geohydrology of Gujarat State*, Ahmedabad: CGWB, 1989.

Regulatory Framework for Water Services in Gujarat, Mumbai: Tata Energy Research Institute.

Report of the Committee on Computerisation of Rural Water Supply and Sanitation Sector, New Delhi: Rajiv Gandhi National Drinking Water Mission, 1995.

Sabarmati River Basin Water Future—technology, policy, institutions, Anand: Indian Rural Management Association, 2001.

Torkil, J.C., J. Fugl, 'Firming up the conceptual basis of IWRM', *Water Resources Development Journal*, Vol. 17, No. 4, 2001.

Water Resources Planning for State of Gujarat, Gandhinagar: Tahal Consultancy Engineers, 1997.

4

INTEGRATED WATER RESOURCES MANAGEMENT IN BANGLADESH: AN ASSESSMENT

ATM Shamsul Huda

Introduction

Scientific approaches to water management in Bangladesh started around the middle of the last century, triggered by the two consecutive floods of 1954 and 1955. Until recently, the management strategy since the 1950s has centred around the development of a 'master plan' that would usually focus on one major contingent theme. The approach to water management was narrowly focussed on either combating natural disasters like floods and cyclones or attaining self-sufficiency in food production through a massive programme of dry-season irrigation. The 1964 Master Plan (EPWAPDA, 1964) and the 1989 Flood Action Plan (World Bank, 1989) represent the former approach, while the 1972 Land and Water Resources Sector Study (IBRD, 1972) and the National Water Plan Phases I (Master Plan Organization [MPO], 1987) and II (MPO, 1991) represent the latter. One common feature of all the plans is that they are investment plans primarily serving the needs of flood control and irrigation and, to a lesser degree, of drainage. All of them lack a comprehensive approach to water as an inherent part of an ecosystem having multiple and competing uses. Despite the imperfections of the plans, they were the only rational basis on which investments in the water sector could be made. Indeed, by the year 2000–2001, an estimated 2.5 billion US dollar worth of investments were made in the sector (Asian Development Bank, 2001). Since there was no framework for integrated water resources planning, the projects executed under those plans did not bear any mark of an integrated approach. At this point of time, Bangladesh does not have any experience of implementing any project or programme that can be labelled as integrated water resources management (IWRM).

During the latter half of the 1990s, a number of developments have taken place in the water sector of Bangladesh that have created the enabling environment for the beginnings of an IWRM. The purpose of this chapter is to carry out a critical assessment of the capacity of Bangladesh to undertake such a process in the near future. We will begin with an attempt to identify the water management issues by an evaluation of the current status of the sector in terms of its linkages with the developmental goals of the government.

Water Management Issues in Bangladesh

An assessment of the status of IWRM in Bangladesh calls for an identification of the issues that need to be integrated. This is done with reference to the concept of IWRM and its general principles as agreed upon in the Dublin and Rio conferences in 1992 and further updated in Bonn in 2001 and in Johannesburg in 2002. In the absence of an agreed definition of IWRM, we proceed with our analysis on the basis of the definition adopted by the Technical Advisory Committee of Global Water Partnership 2000.[1]

IWRM tries to look at water from a systemic perspective and is more concerned than traditional management with the sustainable management of both demand and supply of water. In order to achieve this, there is a need for integration of different constituent elements within both the natural and the human system. In the context of Bangladesh, an IWRM programme will have to seek integration of a number of elements that are discussed here separately for both the systems.

THE NATURAL SYSTEM

The natural system had started off with a steady state of equilibrium of its different constituents. With gradual expansion of human interventions for achieving specific goals, that balance began to be disturbed unwittingly in favour of one or more components (Khan, 1993). This sort of unbalanced and arbitrary action brought about disharmony in the nature. There is a need to restore, as far as practicable, the old equilibrium through an integrated approach to human interventions in the natural system. In the case of Bangladesh, this integration will have to contend with the following issues:

INTEGRATION BETWEEN LAND AND WATER MANAGEMENT

Land and water are the two most valuable resources of the country. The focus on their integration centres round a wide range of interventions in

---

[1] Global Water Partnership, Technical Advisory Committee, 2000.

the country's land and water regime made by diverse interest groups to meet their specific goals. There are innumerable dimensions of this interrelationship and this is not the place to treat the issue exhaustively. For expository purposes, the three most important areas of concern will be touched upon briefly. These are the imperatives of poverty alleviation and food security, resolution of conflicts arising out of competing demand for land use, and protection of environment in the face of interventions in land and water use.

Poverty alleviation and food security depend, to a great extent, on the ability of Bangladesh to carry out a successful programme of intensive agriculture. Given the current disposition of the different factors of production, achievement of this goal seems problematic unless there are significant changes in the current practice of land and water management prevalent in the country.

The total land area of Bangladesh is approximately 14.7 Mha, about 60 per cent of which is under cultivation in a normal year. However, changes in the share of different types of land uses are already noticeable and more changes have been predicted. The net cropped area (NCA)—area cropped once or more during the year, including a small portion of culturable waste and current fallow—declined from 9.3 Mha in 1987 to 8.3 Mha in 1996. Urban settlement and infrastructure development are predicted to further reduce the NCA to about 7.7 Mha by 2025. Against a net cultivable area of 8.3 Mha, 7.56 Mha is suitable for irrigation. Till now, about 4.00 Mha have been brought under irrigation (Water Resources Planning Organization [WARPO], 2000a, b).

In a recent analysis on rice production prospects and demands made by the National Water Management Plan (NWMP), it has been concluded that the current level of rice production in Bangladesh is within one million MT or less of current national demand (WARPO, 2000b). It has also been predicted that to maintain near self-sufficiency, food grains need to grow by around 2 per cent per annum. This increased output will have to be achieved in the face of declining NCA and global climate change.

Optimization of water use for obtaining best results belongs to one side of the equation. But more important for long-term sustainability of a steady growth of productivity is the management of the land as a resource. A number of studies have pointed out that though the major source of crop output expansion has been additional inputs, there is evidence of their decreasing marginal returns. This has since been established in the case of fertilizer (Pagiola, 1995; World Bank, 1995;

Ministry of Agriculture, Government of Bangladesh, 1997). In the case of irrigation also, its marginal contribution is expected to remain more or less constant in the coming ten to fifteen years, and then drop as the irrigation frontier closes (McIntire, 1998). In the face of such prognosis, advances in total factor productivity could be an option along with better water management practices.

The second area of concern in the relationship between land and water has to deal with the conflicts arising out of competing demand for use of land. The Bangladesh National Conservation Strategy has identified six important areas of conflicting land use in the country (Government of Bangladesh and IUCN, 1992). These are: (i) agriculture versus shrimp and capture fisheries; (ii) forests versus shrimp and capture fisheries; (iii) agriculture versus livestock; (iv) agriculture versus human settlements; (v) agriculture versus brickfields and (vi) agriculture versus newly accreted lands on river beds called *chars.*

In general terms, it may be pointed out that earlier, the traditional systems had developed socially sanctioned ways in which these competing demands could be reconciled. Traditional shrimp cultivation was quite compatible with rice cultivation, as was the facility of community grazing field for livestock that protected the crop lands from stray animals (Brandon, 1998). With rapid growth of population, scarcity of land became more acute and the traditional system of accommodation completely broke down. A new system, in conformity with the current realities, has not yet emerged and that is at the root of all these conflicts.

Various studies have shown that it is possible to reconcile the conflicting demands for land use through mutual accommodation. While undertaking this sort of conflict resolution, the prime consideration ought to be finding viable and sustainable production systems. Objective field investigations have revealed that there are areas in which shrimp cultivation has a greater comparative advantage than crops (Rahman, 1990; WARPO, 2000c). Intensive shrimp cultivation may render some fields unfit for rice cultivation; this, however, cannot be a good enough reason for opposing its culture. Local people generally are inimical to shrimp culture due to the selfish and oppressive behaviour of the farm owners who are mostly outsiders. Payment of fair compensation or sharing the profits with all affected landowners and *bargadars* can remove many of the misgivings surrounding this relationship (Habib, 1999).

The Bangladesh government has already incurred huge expenditures in developing infrastructure for flood protection, drainage, and surface water irrigation. Despite some negative consequences, these measures have had a positive impact on agricultural production (Soussan et al., 1998; Datta, 1999). Unfortunately, much of the intended benefits of flood protection and drainage have already been lost due to unplanned human settlements in prime agricultural lands. The Dhaka-Narayanganj-Demra project, the Narayanganj-Narsingdi project, and the Meghna-Dhonagoda projects are prime examples of such encroachment and exploitation. Given the necessity of human settlement faced by an ever-expanding population, it is not possible to deny access to people to settle within a protected area. What is possible is a kind of zoning of project areas that will reserve the best agricultural land for intensive cultivation, directing new settlement to less productive areas (Brammer, 1995).

In the matter of human settlements in the newly accreted *char* lands, existing regulations need to be strictly enforced. Settlement procedures for these lands are clearly enunciated and nobody should settle there till such time the revenue officials have formally allocated the land to eligible persons. However, powerful land-grabbers always pre-empt this process by using money and political connections and send poor people to settle on lands that are not yet suitable for human habitation. They are forced to start cropping on these fragile lands and in the process not only endanger their life but also induce rapid erosion in the emerging *char* (Wilde, 2000).

The third area of concern relates to environmental degradation as a direct consequence of land and water use interventions and is closely linked with intensive agriculture and other land use conflicts. Confronted with the challenge of maintaining a 2 per cent annual rate of growth of food grains with declining availability of land, Bangladesh has to continue with its programme for intensive agriculture. In the process, vital environmental concerns cannot be overlooked. Intensive agriculture somehow has been construed to mean mono-cropping, introduction of high-yielding hybrid crops, and indiscriminate use of chemical fertilizers and pesticides. Such an unbalanced approach has not only reduced the genetic wealth of the country but has also caused land degradation and loss of soil fertility. Common categories of land degradation are erosion, waterlogging, salinity, and depletion of nutrients. Though no firm data is available on the extent of soil degradation in the country, there is some evidence that all these problems do exist in some degree (Brandon, 1998).

Flood control and drainage structures have also played a part in environmental degradation. Countrywide network of this infrastructure have led to the reduction in wetland areas, increased waterlogging due to drainage congestion, and decline in soil fertility due to diminished aquatic vegetation. The worst sufferers, however, have been the fisheries of the country. Flood control embankments disconnected the rivers from the floodplains, cutting migration routes and disturbing the spawning areas of local varieties of fishes (Flood Plan Coordination Organization, 1994). Continued agricultural growth is essential for the overall development of the country but this growth will not be sustainable at the cost of environment as currently being practised. Degraded environment will negate what intensive agriculture seeks to achieve. Land and water use have to be harmonized in a way that will preserve the natural environment as well as optimize the output of all interventions for deriving maximum social benefit.

Integration between Surface Water and Groundwater Management

The division of water resources into surface water and groundwater is artificial and recourse to such a division is made for the purpose of understanding the dynamics of the operation of the hydrological cycle of a particular unit of analysis. Water is essentially a unitary resource and surface water and groundwater are different manifestations of the same resource base (Rogers, 1996). Problems arise when attempts are made to improve the efficiency of the one at the cost of the other in an autonomous mode. In such a situation, only a part of the hydrological cycle is considered, ignoring the entire hydrological balance. These type of interventions, in the long run, will entail a cost on the system that may make the entire operation unsustainable. For long-term sustainability of water resources development, integration between surface water and groundwater management becomes an important issue.

In the context of Bangladesh, this integration is indeed a major issue. The country is rich in water resources and, on an annual basis, supply unquestionably exceeds demand. However, this adequacy is not evenly distributed over space and time and there are seasonal and regional variations. In such a situation, demand management—the use of price, quantitative restrictions, and other devices to limit the demand for water—based on potential sources of supply is crucial.

Given the hydrological conditions of Bangladesh, the desired level of integration will not be easy to achieve. Let us first consider the demands for non-consumptive uses of water. It is to be noted that such demands can only be met by surface water. In exclusive surface water zones, this has also to meet consumptive needs of the local people. But surface water resources appear in such form that accessing them for maximizing benefit become huge and expensive enterprises. In one form, surface water appears as a fugitive resource and moves from one place to the other rather fast. Unless stored or abstracted, it cannot be used. Static resources, on the other hand, stay in particular places for a couple of months and dissipate gradually if not used. Fugitive resources comprise the cross-border flows that enter Bangladesh through the fifty-seven trans-boundary rivers. These flows are augmented by inflows on distributaries from the major rivers and by drainage from local catchments. But these are not enough for meeting the dry-season water demand. The minimum flow during the dry season could be as low as 5000 m^3/s. The NWMP projections beyond the year 2025 call for harnessing the water of the main rivers by constructing dams across the Ganga and the Brahmaputra (WARPO, 2001). However, this strategy has failed to materialize due to the enormity of the enterprise and the associated costs.

Bangladesh is also richly endowed with groundwater resources. In an aggregate sense, the alluvial aquifers of Bangladesh are considered to be among the most productive in the world. Except for the area beneath the metropolitan limits of Dhaka city, adequate rainfall and annual flooding replenish and recharge the aquifers. The NWMP has estimated that reasonable quantity of groundwater is available in 56 per cent of the country. Limited quantity of groundwater is available in other parts also but that can only be abstracted for domestic and municipal use. Even in groundwater-rich areas, there are pockets where such water could be insufficient or technically not feasible to extract (WARPO, 2001). One advantage about groundwater resources is that they continue to remain in place and can be used as convenient; but unlike surface water, local surplus of groundwater cannot be transferred to deficit areas for use, though some of it flows to the adjacent areas.

Despite availability of groundwater in sufficient quantity, there are a number of constraints for its further development. There are certain areas, like coastal areas and offshore islands, where groundwater salinity is a physical constraint to its development. In parts of the north-west and north-central region of the country, low permeability limits recharge and the potential for groundwater exploitation. Another common problem with groundwater development is its susceptibility to over-exploitation. A comparison of potential demands on groundwater availability carried out by the NWMP found that deficits may arise in both the south-west and north-west regions of Bangladesh along the western border where recharge tends to be less (WARPO, 2001). Further expansion of groundwater irrigation development where irrigation is highly developed could lead to serious drawdown of the water table. This may require the use of appropriate technologies for extracting water from much greater depth than are currently being used. Besides having cost implications, such a drive may come in direct conflict with one of the options for arsenic mitigation programme.

The NWMP projections indicate that in the medium term (till the year 2010) the major increase in demand for water will come from agricultural production needs. The demand for water supply for urban and rural areas and for other commercial uses would more than double during that period but in the overall context this increase will be rather small. Meeting this additional requirement for irrigation would not have posed any major problem but for two emerging concerns of significant import. One is the probable effect of global climate change on the water regime of Bangladesh and the other, the implications of arsenic contamination on water resource management of the country (WARPO, 2000b).

The most recent projections about the probable effects of climate change due to global warming indicate drier and warmer winters, wetter monsoons, and rising sea levels (World Bank, 2000). While the probable effects of climate change are yet to manifest themselves, the widespread presence of arsenic in the groundwater of Bangladesh has already been detected and documented (Ahmed, 2000; Department of Public Health Engineering, 1999 and 2000). This is the worst case of arsenic contamination to have ever happened anywhere in the world. This has put to serious jeopardy the achievements made by the country in providing safe drinking water to 97 per cent of its population through sinking of some seven to ten million hand tube wells (HTWs). The main reason for the success of the programme was the low capital and maintenance cost of the HTWs as these are designed to extract water from shallow aquifers. In addition to arsenic contamination, the programme is also under threat of being choked off in some parts of the country where intensive irrigation is lowering the water table. In the coastal areas, such lowering of water table induces salinization, making water unfit for consumption. If no corrective actions are taken, the NWMP has predicted that by 2025, 51 per cent of the population will be living in *thana*s—an administrative unit in between a district and an union (the lowest level administrative unit consisting of ten to twelve villages), composed of six to nine unions—with arsenic in the shallow aquifer, a further 33 per cent in areas with low water tables and a further 7 per cent in *thana*s where the shallow aquifer is saline (WARPO, 2000d).

Field investigations have shown that there is very little arsenic contamination in aquifers below 150–250m. As a longer term, if not a permanent solution, there is a possibility that water abstracted from the deep aquifer will be considered the best option for securing arsenic- and bacteria-free drinking water for the rural people of Bangladesh. This requirement comes in direct conflict with the irrigation needs of drawing water from the same depth due to lowering of water table in large areas of the country. While groundwater is fully replenished every year, its availability at shallow depth is also declining with every passing year.

With full privatization of minor irrigation operations in the early 1990s and surface water irrigation being gradually subjected to operational cost sharing, an environment has now been created for the practice of conjunctive use of water. Despite lack of initiative from the relevant public sector agencies, some spontaneous modest efforts at conjunctive use are noticeable at local levels. In many areas of the country where both surface and groundwater are available, farmers take recourse to conjunctive use by irrigating their land with stored surface water till January and then extract groundwater by tube wells for completing the irrigation cycle. The more water is considered as an economic good and the more publicly provided surface water supplies are brought under the pricing system, the better are the opportunities for instituting conjunctive use of water. In all areas where water is available in both forms, surface water supplies may be developed by local bodies or by relevant public agencies to encourage conjunctive use.

INTEGRATION OF UPSTREAM AND DOWNSTREAM WATER MANAGEMENT

Bangladesh is the lowest riparian of the three mighty rivers entering its borders after traversing through several countries. The Ganga, Brahmaputra, and Meghna (GBM) together constitute one of the largest basins of the world. The country's unique physical setting makes it extremely vulnerable to the hazards of floods and droughts. Since river systems are distinct hydrological units, basin-wide planning for development and management of water resources would be the most logical approach by the co-riparian countries. Regional cooperation is a compelling necessity for the overall management of the catchment and mitigation of floods and droughts, and management of other water-related issues like hydropower generation, water quality control, navigation, and trade and commerce.

In absolute terms, the total quantum of water available in the basin area is enough to meet the entire requirements of all the co-basin countries. Similarly, the terrain in parts of Nepal and India offer excellent sites for storage reservoirs for providing relief from recurring floods downstream along with other beneficial uses. The average annual water flow in the GBM region is estimated to be 1350 billion m^3 (BCM). The annual average water availability per square kilometre in the GBM region is three times the world average. In addition to surface water, the region has an annually replenishable groundwater resource of about 230 BCM. The challenge is to store the monsoon flows and redistribute it over space and time within a mutually agreed framework for the benefit of all the co-riparians (Adhikari et al., 2000).

Studies conducted by Nepal at different levels (master plan, pre-feasibility, and feasibility) have identified about thirty reservoir sites. The total storage capacity of high dam projects identified so far in Nepal is of the order of 82 BCM of live storage that would regulate over 95 per cent of the total annual flow (Bangladesh-Nepal Joint Study Team, 1989). The Brahmaputra Master Plan of India has identified eighteen storage sites in north-eastern India, five of which are classified as large, having a total gross storage capacity of 80 BCM. In the Meghna system, one large storage site has been identified at Tipaimukh with gross storage potential of 15 BCM.

The above proposals relating to the high dams have since been discussed in many forums but have not moved beyond their initial identification stage. Mutual distrust and lack of vision are the two major bottlenecks towards regional cooperation. At the technical level also, the reservoir concept is fraught with many uncertainties. First, even if a few of these reservoirs are built, they will primarily be used for generation of hydropower and the objectives of power generation are in conflict with the objectives of flood control. Secondly, in order to have an impact on peak flows, these reservoirs will have to be of a size for which adequate funding may not be forthcoming. Finally, the environmental implications of such mega projects have not yet been properly studied.

However, with goodwill and mutual understanding, such reservoirs in fact can be so designed and operated as to achieve flood control, irrigation, and power generation objectives. The signing of the Ganga Water-sharing Treaty between India and Bangladesh and that of the Mahakali Treaty between India and Nepal in 1996 are indeed ground-breaking events in the use of international waters in South Asia. This achievement is the fruit of tremendous forbearance and patience shown by the parties concerned. At this stage, the concept of basin-wide planning does not exist but recent events have raised the hope that such cooperation may be achievable in the not too distant future.

The other area for upstream and downstream cooperation is in the field of flood forecasting and warning. Already there are working arrangements between India and Bangladesh for transmission of flood-related data. However, present arrangements allow the concerned agency of the Bangladesh government to forecast with a lead time of forty eight hours for the central part of Bangladesh and only four hours for some areas near the border. It is possible to provide more reliable forecasts with more lead time if data is made available by the upper riparian countries from additional upstream points and more frequently. As far as India is concerned, this can be achieved easily to the extent that both India and Bangladesh use similar technologies for data observation and transmission. Similar arrangements also need to be worked out with Nepal and Bhutan.

INTEGRATION BETWEEN FLOOD AND DROUGHT MANAGEMENT

Floods have been a regular phenomenon in Bangladesh since time immemorial. Villagers and farmers make a clear distinction between flooding and floods and it is useful to maintain this distinction in any flood analysis. About 27 per cent of the land in Bangladesh goes under water every year and this is what local people call normal flooding. For generations, people living in these areas have adjusted their way of life to the hydrological cycle and the annual event is taken for granted as part of their lives. This symbiotic relationship between people and nature dramatically changes with the onrush of excessive water than in normal years, which interferes with their life and living. According to Hugh Brammer, 'Floods represent abnormal flooding, when floodwater rises earlier, higher, more quickly or later' than expected (Brammer, 1995). The intensity and duration of floods vary from place to place and year to year. Floods cover 37 per cent of the country every ten years and 60 per cent every 100 years.

Floods, as distinct from flooding, become major public concern due to its adverse effects on the economy and affected people. Over the years, the extent of flood damages for the same level and duration of floods occurring in the past have increased manifold as a result of rapid growth of population, urbanization, and economic development.

While there is too much water during the wet season, there is sometimes too little during the dry period. Generally, during the dry season, spanning from November to May, the amount of rainfall in the entire GBM basin is very small. This is also the time when significant quantity of water is diverted in the upper reaches of the basin. All these result in drastic reduction in the discharge of the major rivers entering

Bangladesh, drying out of waterbodies, falling water levels, and salinity intrusion. Due to intensification of agriculture, water demand for irrigation is quite high and there is a pressure for more extraction of groundwater. In areas with poor access to surface water and groundwater resources, drought causes real hardship to the affected people. Drought, however, does not get that much attention of the concerned people for the reason that drought losses are not easily identified and measured. A comparative analysis of the respective losses in rice production suffered by flood and drought during the period 1969–70 and 1983–4 revealed that drought is more damaging to aggregate production than floods (World Bank, 1998). There were severe droughts in Bangladesh in 1979, 1981, 1982, and 1989, and between November 1998 and April 1999, there was a period of 150 days with almost no rain in Bangladesh.

Like floods, the effects of droughts are more intensely felt these days for the reason that irrigated *boro*—a rice crop planted under irrigation during the dry season from December to March and harvested from April to June—has become the major rice crop. In times of serious water stress, the farmers suck up whatever little water is available in the waterbodies by using low lift pumps. This deprives the waterbodies of their in-stream needs (includes such non-consumptive uses as water for navigation, fisheries, and salinity and pollution content) for fisheries and the environment. The shallow tube wells are also lowered to reach water for extraction. This drawing down of water has a corresponding impact in lowering the table for village HTWs installed for supply of drinking water. The incidence of drought thus sets in motion a chain reaction that not only affects agricultural production but also navigation, drinking water supply, environment, rural health and sanitation, and general welfare of the people.

Until recently, flood management has been approached by looking at flood as an independent variable and not as a component within the framework of an overall water resources management. Construction of embankments was the preferred method for preventing floodwater from entering into designated areas. By the end of the year 2000, about 9000 kilometres of embankments were constructed all over the country. Different evaluations as to the efficacy of embankments present a mixed picture. There is unmistakable evidence that more successful embankment projects increased agricultural productivity and created noticeable non-farm employment and, in some areas, there were marked improvements in the health and nutrition status of the local people (WARPO, 2000e). But the incidence of failures was more numerous than the success stories.

Following very intensive debates generated by the Flood Action Plan of the 1990s, there is now a broad consensus that full protection from flood is neither possible nor desirable. Major cities, towns, commercial centres, and places of historical importance need to be protected from flood and erosion by structural measures of required standard. For the rural communities, flood mitigation measures should consist of a strong programme of flood proofing supported by a dependable flood forecasting and warning system. Flood proofing was very much a part of the life-style of the Bangladesh people: this has to be revived. Once homesteads are raised, water supply and sanitation can easily be flood proofed. However, this is an activity that largely belongs to the domain of private citizens and they must be motivated to follow the initiative taken by a few of the NGOs.

While it is possible to suggest a few methods for coping with the adverse impacts of catastrophic floods, it is very difficult to come up with any pragmatic suggestions for drought management that can be implemented easily. The best option for resolving this problem is the augmentation of the river flows during the dry season. As has been discussed already, this is not going to happen soon and one has to look for other solutions.

The national water policy envisages future groundwater development by both the public and the private sectors and does not rule out the possibility of regulation when necessary (Ministry of Water Resources, Government of Bangladesh, 1999). The Ministry of Agriculture itself considers regulation as one option for managing drought (Ministry of Agriculture, Government of Bangladesh, 1999). The NWMP examined the possibility of preventing the use of LLPs in designated zones as well as limiting abstraction by non-essential major users during times of drought. However, on grounds of difficulty of enforcement, it was not considered a viable option (WARPO, 2000e). Improvements in irrigation efficiency are often suggested as a way to mitigate water shortages. At an individual level, such improvements benefit the farmer who has to pay less for water, if it is priced. It is also a desirable national goal for bringing down the cost of production. Its contribution to shortage mitigation, however, is insignificant. Unused or wasted water is recycled by users downstream. Any improvement in efficiency by upstream users will require the people downstream to look for alternative sources of water (Siddiqi, 1998).

In these circumstances, the only option left open is demand management. In countries where water is treated both as an economic and a public good, water is priced for certain categories of deliveries and a water market is fully operational. Demand management through water pricing then becomes a mechanism for controlling scarcity. In Bangladesh, water is still looked upon as a bounty of nature and as a public good. Though a rudimentary market for tube-well irrigation has been developing, the time is not yet ripe for a full-scale market operation. The best that can function here is a mixed system where tube well and pumped irrigation and piped drinking water supply in the major cities and secondary towns are subjected to an evolving water market operations, leaving other operations to be administered by the state.

Integration between Water Quality and Quantity

The quality of water is dependent, in various ways, on the quantity of water available at particular time and space on a regular basis. Both excess and scarcity of water have harmful effects on the aquatic environment. The aim of successful IWRM is to achieve a level of water availability that would ensure reasonable quality of water. For Bangladesh, the issue becomes all the more complex due to its location as the lowest riparian and a total lack of control over 97 per cent of the catchment area.

Over the years, the quality of surface water in the region has deteriorated to an alarming proportion. Bangladesh has to bear the burnt of upstream pollution in the form of sediment load, industrial effluents, agro-chemicals, and domestic wastes that pass through her river network. To this are added the locally generated pollutants of similar type as well as faecal pollution due to widespread lack of sanitation facilities. Village ponds, exclusively used in the rural areas for various uses including water for cooking, also get contaminated during the monsoon as inflows from the contaminated area increase. Too much water in an area with poor drainage can cause waterlogging, triggering long-term changes in the aquatic environment. Sediment load transported from the upstream renders water unsuitable for certain uses without treatment and causes bed levels in the rivers to rise.

The adverse effects of some of the pollutants carried by water can be greatly mitigated by the water regime itself. Over 9 per cent of Bangladesh's. area are covered by rivers, estuaries, and wetlands. With sufficient quantity of in-stream water, they can perform such vital functions as: (i) retention and regulation of flood flows, (ii) transport and retention of sedimentation, (iii) dilution, dispersion, and assimilation of pollutants, and (iv) sustaining aquatic and terrestrial ecosystems and, indirectly, land-based economic activities, including farming. Unfortunately, water is not available in the quantity required during the dry season for quality maintenance. Additionally, there are the problems of salinity and arsenic in the groundwater. Scarcity of water during the dry season has increased the salinity level generally in the coastal areas and in the Sunderbans in the south-west of Bangladesh in particular.

A long-term solution to the problem of water quality depends on adequate in-stream water in the rivers and waterbodies during the dry season and this can be accomplished only through regional cooperation on water sharing and management. It may take considerable effort and time for Bangladesh to work out joint plans for different river basins with other co-riparian countries. In the meantime, efforts must be made to ensure good water quality by taking appropriate measures as are feasible within its borders.

INTEGRATION BETWEEN SCALES

Natural water resources system constitutes a distinct hydrological unit and, from that perspective, river basins are very appropriate units for operational management. However, it is impractical to expect that all interventions in the system will be made centrally by a super agency to protect its integrity. Issues pertaining to water resources are faced by people at different levels—regional, national, district, village, and household—in Bangladesh. It is imperative for an IWRM to define hydrological boundaries in such a manner that it is able to meet the requirements of both system integrity and decentralized water management within a system framework.

The task of achieving integration between units of different scales would involve the following actions:

(i) Dividing each hydrological unit into sub-regional and local units within the broad framework of the larger unit so as to make it feasible to make the necessary interventions in the water regime to meet unit demands without any major disruption to the system;

(ii) Defining the parameters of water resources planning in an integrated manner and providing the framework for developing sub-regional and local-level plans;

(iii) Developing mechanisms for ventilation of local and sub-regional points of view on water resources management at higher levels, particularly at national and regional levels;

(iv) Encouraging multiple use of water resources to identify trade-offs among its different uses for the purposes of conservation and conflict resolution;

(v) Establishing and maintaining a national water resources database; and

(vi) Supporting continuous research on the natural system and its linkages with other natural resources and the physical system.

Integration between scales means more than meeting local water needs. The lower units are the building blocks of the hydrological units that affect them or are affected by them. This reinforces the importance of a very strong hierarchy of hydrological units in the natural system, for it is ultimately the users who determine the value of a natural system in terms of its utility to mankind.

HUMAN SYSTEM

The natural system of water resources exists in the environment as a bounty of nature ready to be exploited for service to the teeming millions of people. Whether the concerned people are able to do this job in a socially, economically, and environmentally acceptable manner is a relevant question in an IWRM context. A technical guide on the subject has divided human activities centring around water into two broad categories: in the first category belong interventions that directly affect stream flows, water quality, and groundwater, while in the other category belong interventions in the catchment basins (Guy Le Moigne et al., 1994). The spectre of drought, polluted water bodies, desertification, salinization, declining fish stocks, rising sea levels, and decaying ecosystem are not so much the result of scarcity of water as against failures of different systems to manage it properly. The key issue for IWRM is to develop an appropriate human system that is responsive to people's needs and is generally capable of delivering the intended services without jeopardizing the sustainability of the natural resources. In other words, it must be able to work with the natural system to achieve the national goals. At a minimum, the human system should address the following issues:

MAINSTREAMING WATER IN THE NATIONAL ECONOMY

Since time immemorial, Bangladesh has been blessed with abundant water. It is only in very recent times that she has been feeling a little stressed for water during the dry season. Easy availability of water has led people to believe that water is a free gift of nature and that it can be used in any manner they like. This freewheeling style of water management did not cause any major problem as demand for water was moderate relative to its availability. Government approach to water management also reflected this popular perception. Until very recently, management of water-related calamities and accelerated rice production through irrigated agriculture were the two dominating themes in water management by the government.

Water use has been coterminous, positively, with agriculture development and, negatively, with protection from flood and cyclones. For a variety of reasons, this situation has been changing rapidly. The time has come to rescue water from these narrow confines and focus its true role in the overall development of the economy of Bangladesh through an integrated approach to its management and use.

Mainstreaming water in the national economy of Bangladesh is beset by the legacy of treating it in a fragmented manner by the different agencies. There has been a continuous failure on their part to work together under a lead agency and project to the policy makers a comprehensive vision on use and management of water. Water is a key resource in poverty alleviation programmes of the country but none of the major agencies in the water sector have so far successfully articulated the essential linkages between water and poverty. Consequently, requests for allocation of requisite funds for these agencies are easily ignored by the government on grounds of priority in other sectors. Resource allocation to the water sector has been declining over the past decade.

The top management of the concerned ministries and the agencies can perhaps change this situation. An enabling environment has been created and the supporting documents are ready. They have only to use them skilfully. The ministries connected with the management of one or the other aspect of water resources management have all formally announced their policy indicating their preferred goals. Many of them have also developed follow-up action plans. A recent review of these policies have reported that 'in most cases the policy framework across related sectors is not only mutually compatible, but often mutually reinforcing' (Ministry of Agriculture, Government of Bangladesh et al., 2002). Another earlier review had also expressed similar views (PDO-ICZM, 2001). Based on those policies, a National Water Management Plan (NWMP) has already been prepared. The main objective of this plan is poverty alleviation, and the linkages between water and poverty are well articulated. It is time now for action. Putting water in its rightful position can only ensure sustained attention of the government to the sector which is a prerequisite for IWRM.

INTEGRATING DIFFERENT USES OF WATER

Some thirty-five public agencies are connected with water as a basic resource to fulfil their organizational mandates. Public management of resources in Bangladesh is generally highly departmentalized, more so in the case of management of water resources. Water management is riddled with conflicts between sub-sectors due to disjointed policies on different aspects of water use. Lack of a coherent policy and planning on the use of water has brought the situation to such a pass that some of the agencies were found to have been working at cross-purposes. Now that the relevant ministries have declared their sub-sectoral policies and these have generally been found to be compatible with each other, they should start working towards harmonizing the remaining inconsistencies.

Integration of different uses of water is needed not just for minimizing inter-sectoral conflicts but also for ensuring equity and access to disadvantaged groups, particularly women. During the dry season, there is water scarcity in certain parts of Bangladesh. But in other parts where water may be available in reasonable quantity in an absolute sense, a condition of scarcity is created for some by different kinds of social, economic, technical or institutional barriers which limit access to water. The quantum of water available in a particular jurisdiction may be good enough to meet everyone's demand. However, due to a distorted distribution mechanism, there is more than enough water for some while very little for many others. With an ever-increasing population and rapid urbanization, incidence of scarcity owing to distorted distribution will increase unless urgent measures are taken to remedy the situation. This kind of man-made scarcity gives rise to social conflicts that sometimes can turn violent. Such conflicts appear in many forms. These are mostly local, between the same type of users in one place (fishers vying for fishing rights in the same waterbody), between different types of users in the same place (shrimp versus rice cultivation in the same field), and between different localities (water control structures upstream denying adequate water to people downstream). The seed of discord is inherent in the nature of water that has multiple uses. These uses are not compatible with each other and therefore there is a need to find the complementarities and trade-offs for minimizing the conflicts.

INTEGRATION BETWEEN PUBLIC AND PRIVATE SECTORS

The delineation of the respective jurisdictions of the public and the private sectors in water resources management is an important IWRM issue. Such a determination would be conditioned by the imperative of the state to provide essential services to its citizens that cannot be left to the whims of the market. This imperative has two broad dimensions: those water-related activities that bear the character of public good and use of water in waterbodies that are generally treated as common property in the absence of a firm legal position on their status.

With the expansion of the economy of Bangladesh, the demand for water will increase and water rights and allocation issues will need to be resolved. Since water resources in the open waterbodies are treated as common property in Bangladesh, effective allocation of this resource between uses and users cannot be left to the market. Similarly, public investment in the water sector needs to continue towards creation of public goods, addressing specific market failures, or for protecting particular community interests. The state has to be involved in the twin role of major investor in water resources development as well as an effective mediator in deciding allocation rules and enforcing them. This does not mean that the government should be directly involved in managing the programmes and enforcing the regulations down to the individual users. Government should confine its role to broad policy-making and execution of major projects of a public goods nature. Management of water projects may gradually be transferred to the private sector. Initially, public sector institutions may institute joint management system for the large projects with the ultimate aim of full management transfer to the private sector or the beneficiary group while management of the medium and small-scale projects should be fully handed over to the beneficiaries.

HORIZONTAL AND VERTICAL INTEGRATION AMONG RELEVANT INSTITUTIONS

In modern-day governance, integration of efforts through coordination and cooperation among a number of institutions is an organizational imperative from which there is no escape. Not only are the governmental operations highly interconnected, a large majority of the problems are also multidimensional in character, defying the ability of one single ministry to resolve them. These problems can only be resolved by taking recourse to the knowledge, skills, and technology available in a number of ministries and agencies in close cooperation with the private sector, NGOs, and relevant members of the communities.

Narrow departmentalism is a major stumbling block towards achieving the desired level of integration which is engendered and nurtured by subsectoral planning and execution of projects by the concerned ministries and the agencies under their control. Growing awareness about the integrated use and management of natural resources has led a few agencies to develop sectoral action plans involving concerned ministries and agencies. National Environment Management Plan (1995) and National Water Management Plan (2002) are the outputs of joint efforts by all stakeholders. The integrated coastal zone management programme is also being prepared through the involvement of all concerned stakeholders.

Incidentally, these efforts are confined in two ministries only, the Ministry of Water Resources and the Ministry of Environment and Forests. Though modest, the thrust of the efforts are significant to the extent it seeks to improve horizontal coordination at the planning stage first. Agreement of all relevant government agencies, the private sector, and the concerned public on an action plan is a major first step towards integrated planning and use of natural resources.

The second element in inter-organizational integration relates to vertical integration. Administration in Bangladesh is not only departmentalized but also highly centralized. The country is administered through a unitary form of government that extends up to the remotest corners of the villages. Ministries are responsible for policy formulation and supervision and monitoring of the activities of the agencies under their control who actually execute those policies and directions in the field. In the absence of an operational local government system, field offices of these agencies are supposed to deliver the services to the people. The quality, speed, and frequency of these deliveries are impeded by lack of authority at the local level. Decentralization, delegation, and deconcentration are proven methods for improving vertical integration but very rarely taken recourse to. It is hoped that horizontal integration over time may facilitate vertical integration of the line agencies when they would be called upon to implement time-bound action plans.

INTEGRATION BETWEEN SERVICE PROVIDERS AND BENEFICIARIES

Until recently, development activities in Bangladesh used to be bureaucratically determined and executed. The water sector was also no exception. Through experience, the water sector institutions have learnt about the immense benefits of beneficiary involvement in water resources planning and management. Beneficiary involvement in project design and management increases the efficiency of project implementation and the effectiveness of project sustainability. Case studies (Soussan, 1998) have revealed the complexity of the rural society of Bangladesh and its water management. Management of water resources takes many forms and is regulated by many institutions including customary rights, traditions, and social norms as well as more formal types of organization. The participants showed a long and well-developed tradition of water resources management. Many of the local initiatives identified were cheaper and more appropriate than external interventions. These were also well integrated with the local institutions.

Participant involvement in all stages of project cycle is also essential in projects where participants are required to share part of the cost or where their support will be required for cost recovery.

Above all, participation is a kind of empowerment. It means self-development in terms of greater realization of human potentials. It assigns to the individual the role not of subject but of actor who defines the goals, controls the resources, and directs the processes affecting his life. It is indeed a substantial decentralization of decision making. It is not an expert-dominated, non-consultative mode of central decision making simply replicated at the lower level. Decision making must truly be returned to the people who have both the capacity and the right to inject into the process the richness of their values and needs.

While the benefits of participatory approach to water management are understood by the water sector institutions, the process for its institutionalization is rather slow. Developing beneficiary organizations is done in a routine manner while these need a lot of caring and a high level of commitment. Realization of the share of O&M cost from the beneficiaries has also remained much below the target that is already set at a very low level. Two steps are necessary for further development in the area. Real management of small-and medium-scale water projects ought to be transferred to the beneficiaries after making them fully functional. They must also be allowed to retain the money collected as O&M cost with freedom to use it for the operation of the project according to their own priorities.

INTEGRATING THE ROLE OF WOMEN IN WATER MANAGEMENT

Women in Bangladesh are subject to discrimination despite the fact that, both in urban and rural areas, they are the custodians of family health, nutrition, and welfare as well as managers of domestic water supply and sanitation within the home (Paul, 1999). Rarely do they have the opportunity to express their opinion on water management issues that vitally affect their lives. Decisions on siting public water facilities, management system, and cost recovery are all taken by male members of the community who do not understand the particular needs of women and children. The NWP has given clear guidelines for integration of women in the management of water resources. The Second and Third World Water Forums also expressed their commitment to enabling women to participate in all aspects of planning and management of water resources. The two major multilateral donor agencies, the World Bank and the Asian Development Bank (ADB), have also issued operational guidelines that call for developing programmes that are equitably beneficial to both men and women. The ADB guidelines particularly target gender issues of poverty and access to basic services. The climate for integration of women in water management is favourable and can be acted upon with full international support. There are many options to start the process. One way is to ensure participation of women through the entire project cycle, particularly of drinking water supply and sanitation projects, and transfer responsibility of operation and management to them. Another approach could be to encourage them to opt for employment in the sector and to give preference to women in specific jobs. In Bangladesh, the number of households headed by women is increasing, either due to outmigration of able-bodied male members or an increase in proportion of divorced or widowed women. There is thus an urgent need to integrate them in the mainstream for the sake of establishing equity and reducing poverty.

INTEGRATION BETWEEN THE ACTION PLAN AND ITS EXECUTION

Action plans are to be implemented within a fixed timeframe and to that extent these are to be looked upon as phased execution within the framework of a long-term plan. It is wrong to assume that planning ends with programme development. In an effective IWRM programme, planning is a continuous process in as much as water represents a dynamic resource and changes are occurring constantly. There is thus a need to update such plans at regular intervals through a consultative process involving all stakeholders.

While the integrated action plan provides a broad macro-level framework of action, a major part of the plan will be executed at the regional and sub-regional levels. The integrity of integrated planning will critically hinge upon the capacity at these levels to prepare and implement sub-regional and local water management plans in conformance with the macro-level action plan. For this purpose it may be necessary to set-up project appraisal committees consisting of local representatives of all concerned agencies, NGOs, private sector institutions, and community organizations. These committees will keep a close watch on conformance. There will be occasional conflicts and disputes and for resolving them suitable conflict resolution mechanisms need to be developed.

INTEGRATION OF NATIONAL NEEDS WITH DONORS' INTERESTS

The NWMP has developed an ambitious programme for execution till the year 2025 with a cost estimate of more than US$ 18 billion. It will not be possible for the Bangladesh government to execute this programme without the active support of the donor community. Unfortunately, in recent years, obtaining the agreement of donors on some activities considered vital by the government has become extremely difficult. Even on investment proposals that are finally ironed out, a number of conditionalities are imposed on the government that create unintended implementation difficulties. These conditionalities are sometimes generic in nature and at other times sector specific. Generic conditions relate mostly to such issues as macro-economic fundamentals, institutional reforms, and governance. Sector-specific conditions relate to such issues as restructuring of existing organizations, cost recovery, downsizing of organizations, and amendment of existing laws or promulgating new ones. These are issues that can be discussed and it is possible to reach an agreement on how to address them and under what timeframe. The donors also demonstrate their strong predisposition for or against certain types of investments irrespective of the views of the government on them. This is an area where discussion hardly helps to the extent that the concerned donor has already a view on the issue and demonstrates extreme rigidity in its approach. It is in these areas that hardly any progress is ever made. The NWMP contains a few big infrastructure development projects like the Ganga Barrage project and river bank erosion control projects that can only be implemented with the financial assistance of major donors like the World Bank. In the past, the World Bank has reacted negatively, for example, on the Ganga Barrage even when quite a number of bilateral donors were willing to consider it on merit. In the context of IWRM, Bangladesh will need to carry out some capital-intensive projects that are vital for ensuring its water security. To achieve this, donors will have to show more flexibility and accommodation than has hitherto been exhibited by some of them.

INTERNATIONAL COORDINATION

Prospects of water resources development in the eastern Himalayan region have been very intensely studied. A number of recent studies (Ahmad et al., 1994; Biswas and Hashimoto, 1996; Ahmad et al., 2001; Biswas and Uitto, 2001) have shown that integrated river basin planning can facilitate the creation of the necessary water control structures for both augmentation of flows during the dry season and storage of water for flood moderation during the monsoon. In addition to removing the trap of too much water in one season and too little in the other, such measures will facilitate inter-modal transport development, hydropower development and exchange, further expansion of irrigated agriculture, fisheries development, and an overall improvement of the water ecosystem. Regional countries are aware of the benefits of integrated basin planning and there is continuous dialogue between them both formally and informally at different levels. Efforts at government level have been supplemented by a group of dedicated water experts whose contributions have significantly advanced regional awareness about the issues. At this stage, integrated river basin planning does not seem to be a possibility in the near future, though that is the goal for which all countries in the region should strive. Under such circumstances, Bangladesh has to live with the second best option of trying to manage its water resources as best as she can till basin planning becomes a reality.

Bangladesh and the Framework for IWRM

An IWRM programme can only be launched when the necessary conditions are sufficiently met or are in the process of being met. These conditions are represented by the following building blocks: an enabling environment, a sound basis for planning and prioritization, and a strong and an effective institutional base to carry forward the national goals of an IWRM. The different conditions subsumed by these building blocks are discussed along with a review of the status of Bangladesh in terms of its capacity to undertake such a programme.

CREATING AN ENABLING ENVIRONMENT

For almost the entire period of the last century, water management in Bangladesh has rested firmly in the hands of the public sector. Due to fragmented and departmentalized structure of government and bureaucratic management of water, a number of maladies have crept into the existing system. The most notable of these, from the point of view of externalities, is similar to those happening in many other countries of the region. These are identified as 'fragmented public investment programming, and sector management that have failed to take account of the interdependencies among agencies, jurisdictions and sectors' (World Bank, 1993). A second set of maladies relates to the bureaucratic mode of operation that has often hindered public participation and engendered a culture of lack of transparency and accountability. Yet another set accounts for the neglect of water quality, health, and environmental concerns. Distorted policies, misguided investments, and weak institutions have led to the perpetuation of a vicious cycle around water that needs to be broken. An enabling environment must be created to allow all elements of society to perform their roles in water management.

NATIONAL WATER POLICY

The first essential step towards creating an enabling environment is the formulation and declaration of a national water policy. Bangladesh did not attempt to formulate any water policy since her independence in 1971. It was only in 1999 that the government approved the National Water Policy (NWP) and has since widely circulated it to all concerned (Ministry of Water Resources, Government of Bangladesh, 1999). This document was under preparation for three long years and underwent several revisions. It was drafted by government officials and national experts after due consultations with all concerned public agencies, NGOs, members of the civil society, and the public at large. In its formulation, extreme care was taken to protect, as far as possible, all interests by ensuring a pragmatic and neutral perspective. The NWP provides a comprehensive policy framework for dealing with such issues as river basin planning, water rights and allocation, delineation of public and private domains, water supply and sanitation, preservation of the natural environment, and the developmental concerns of fisheries, navigation, and agriculture. Subsuming these sub-sectoral issues, the policy also provides clear guidance on its disposition towards water as an economic good, water pricing, fuller participation by stakeholders, decentralized management, and delivery structures. The policy has also very clear and unequivocal views on regulations, incentives, public investment plans and environmental protection, and on the interlinkages among them. It declares clearly and unequivocally the intention of the government that all necessary means and measures will be taken to manage the water resources of the country in a comprehensive, integrated, equitable, and environmentally sustainable manner.

NATIONAL WATER MANAGEMENT PLAN

The principal instrument in carrying out an IWRM programme is an action plan that is sometimes designated as national water plan or national water management plan. After two failed attempts in 1987 and 1991, a team of government officials and local and expatriate consultants have succeeded in producing a final version of a National Water Management Plan (NWMP) that has generally been accepted by all stakeholders including the government. The NWMP is different from the previous master plans on two counts. While the master plans have been more concerned with the use of water in one or more sectors, the NWMP has looked upon water as part of a natural ecosystem and has endeavoured to maintain its integrity from a holistic perspective. The other important distinction lies in the approach of the NWMP about the methodology of its implementation. The plan addresses water issues in such a manner that it fosters inter-organizational cooperation and provides details of how, when, and by whom the directives of the NWP would be implemented. Unlike a master plan that is normally sector-oriented, the NWMP also provides a broad framework within which other agencies and organizations at different spatial levels have the freedom to work. As a comprehensive long-term water management plan, it responds to national goals and objectives and has developed schedules according to appropriate time periods. The plan provides for regional balance, subject to the requirements of overall economic and social development and to protecting the natural environment.

No plan can be a one-shot affair and provisions have rightly been made in the NWMP for its continuous review and updating every five years. WARPO has been given this responsibility and for this reason steps are being taken to augment the capacity of the organization. Within a time horizon extending up to 2025, the NWMP provides a firm plan for the next five years (2002–7) that is also coterminous with the duration of the upcoming sixth Five-year Plan. It also provides an indicative plan for the subsequent five years and a perspective plan to 2025. The process of the development of the NWMP involved very intensive interaction of the concerned agencies and other stakeholders and it is expected that sector agencies of the government and the local bodies will prepare and implement sub-regional and local plans in conformance with the framework provided in the NWMP.

LEGISLATION

Many of the management approaches required in an IWRM system run counter to existing administrative and management practices. The NWP dictums relating to decentralized delivery systems, more private sector involvement in water management, water pricing, and management and transfer of ownership of water sector projects to the private sector or the beneficiaries would require drastic revisions of existing laws or regulations and promulgating new ones in their place.

The NWP duly recognizes the imperative of a sound legal system for the effective implementation of the policy. It assigned the initial task of reviewing and identifying the needs of promulgating new laws/regulation or suitably amending the existing ones to the NWMP. The latter has identified four broad areas where the envisaged activity will need to be carried out at the earliest opportunity (WARPO, 2000e). These are:

(i) *Resource ownership, allocation and rights* belong to one area that will influence the determination of the priorities in the use of water, particularly under conditions of scarcity. In Bangladesh, the state is the owner of all water resources and sets the priority of water use between the sectors. But for better transparency and greater accountability, there should be appropriate rules and regulations on these subjects. It has been speculated that some direct regulatory instruments may emerge over time for establishing water rights, abstraction or discharge of water from any source, land-use planning, and utility regulation so far as these would relate to private individuals or groups (Choudhry, 2001).

(ii) The second area has to deal with the setting up of *standards*. These are important for protecting user rights and controlling many types of abuses of water such as water pollution, wastage, and the degradation of the wetlands. Standards are also important for ensuring the quality of different kind of services delivered in the water sector.

(iii) The third area broadly covers the contentious issue of *delineating the respective domains of the public and private sectors in water management*. Water management in Bangladesh is largely controlled by public sector institutions. However, starting in the mid-1980s, the Bangladesh government has by now successfully privatized the ownership of the minor irrigation equipment with the exception of a couple of special projects. The NWP has set the target of transferring the ownership and management of small-scale flood control and drainage (FCD) projects and flood control, drainage, and irrigation (FCDI) projects up to 1000 hectare to the local bodies. The government also plans to transfer management of medium-scale projects to the beneficiaries while instituting joint management of large projects by the Bangladesh Water Development Board (BWDB) and the beneficiaries. The NWP also wants the public sector institutions to use private providers of specific water resources services in carrying out their mandate. The BWDB Act has been thoroughly overhauled in 2000 to accommodate the proposed policy changes. But much more needs to be done to create the enabling environment for the introduction of IWRM in Bangladesh.

(iv) The last and an emerging new area that will need special attention is *financial management of water resource infrastructures*. For too long, water has been looked upon as an abundant bounty of nature to be used for free. In their drive for rapid expansion of irrigated agriculture, successive governments in Bangladesh also downplayed the financial aspects of pubic sector investments in the water sector. Barring the large water supply projects in the cities and big towns, water users were not called upon to bear part of the cost of investment or even pay for operation and maintenance costs. This attitude towards cost recovery or cost sharing has been changing gradually. Nobody in Bangladesh now seriously questions the principle that in the long term, the public sector water agencies should become financially self-supporting with full authority to charge, collect, and make direct use of the fund so collected. New laws or regulations will be needed to enable the authorities to charge the beneficiaries for the services provided by the water sector institutions. Similarly, there should be laws to penalize the offenders responsible for degrading the quality of water. Finally, the existing laws or regulations do not permit the water service organizations to retain the collected fund at the project level. These are all deposited to government treasury, purportedly to be recycled to the projects through the budgetary mechanism. This is not a very transparent and visible system. Beneficiaries want to see that the money paid by them is directly used for the operation and maintenance of the project. The laws or regulations should be so amended as to allow the project authorities to retain the cost recovery/cost-sharing fund for use in the project.

The NWP has provided for enactment of a national water code that will revise and consolidate the laws governing ownership, development, appropriation, utilization, conservation, and protection of water resources. A first working draft of the code has already been prepared and is now being internally examined by the concerned agencies and ministries of the government. The promulgation of the code will take some time as finalizing legal matters entail lengthy procedures in Bangladesh. As and when this is done, it will fill a large vacuum in the legislative framework for the management of water resources.

PLANNING AND PRIORITIZATION

Water has many competitive uses and in the matter of its use, decision makers will have to look for complementarities and trade-offs for minimizing conflicts and ensuring optimal utilization. In making judgements about programme contents and allocation of resources among competing interests, it is essential that different options are prepared on the basis of latest data collected from all relevant sectors. For this reason, this building block of IWRM lays special importance on *information system, assessment of water resources,* and *management procedures.* These three elements will be discussed briefly to put the issues in perspective.

INFORMATION SYSTEM

A sound water resources information system is necessary in estimating the quantity and quality of water as well as the current and prospective water use and demand patterns. Uninformed or distorted decision making will harm the long-term interests of the country. For these reasons, the NWP has emphasized the need for the quick development of a reliable information system and has assigned the responsibility to the WARPO for development of a central database and management information system to be called the National Water Resources Database (NWRD).

The NWRD has since been set-up as an operating unit of the WARPO. However, this effort is hampered by at least three major obstacles in achieving its goals. More than twenty-four agencies are involved in the collection, assimilation, and dissemination of water-related data, resulting in unnecessary duplication of the data so collected, inconsistencies in similar data collected by different agencies, and poor quality data. Planners spend much time and money to correct and assemble data sets but the results are seldom fed back to the collecting agency. Secondly, except for a few items, there is no periodic updating of data once collected under donor funding. Recently, funding has been available for development of powerful computer programmes and geographical information system but far less funding has gone into the collection and monitoring of high-quality spatial and temporal data. It has been alleged that models developed in 1997 were still using data collected for the models produced in 1984 (Rasheed, 1998). This situation is not any different today. And finally, smooth flow of information is hampered by many administrative barriers and, in a few cases, there is a total denial of access on grounds of secrecy.

WARPO has been negotiating in earnest with all the concerned agencies to develop an agreed guideline that will enunciate the principles of common standard, protocols, data sharing, and cooperation and charging. A task force constituted under the leadership of the Bangladesh Bureau of Statistics has made good progress and will come up with a methodology for better information management shortly.

ASSESSMENT OF WATER RESOURCES

A number of analytical tools are already available and a few are in the process of being developed for water resources assessment. Among these, we will briefly touch upon mathematical modelling, blue accounting, and environmental impact assessment.

Mathematical Modelling: Mathematical modelling has emerged as a very powerful tool for the management of water resources in Bangladesh. Its development began in 1986 with the launching of a surface water modelling programme sponsored by the UNDP. Initially, it concentrated on development of six regional models and the general model. The six regional models were stand-alone models of over 12,000 km of river and associated floodplains covering around 90 per cent of the surface water system in Bangladesh.

The first attempt at scientific assessment of water resources of Bangladesh was made in 1991 by the Master Plan Organization (MPO) set-up under the Ministry of Water Resources to develop a master plan for water for the country. This assessment was based on a number of mathematical and simulation models. Dividing the country into five hydrological units that consisted of sixty planning areas, the MPO developed a water balance model and derived the quantum of surface water availability for the 173 catchments falling within those jurisdictions.

For assessment of groundwater potentials, the MPO considered four models that yielded a range of available groundwater values within respective confidence bands. The MPO estimates looked highly conservative and on the lower side vis-à-vis the estimates made by the National Minor Irrigation Development Programme (NMIDP). This made both the estimates suspect. The NWMP bore the mantle of coming up with generally acceptable estimates. For this purpose, the general model and the four regional models were run by the NWMP over a twenty-five-year period to determine the statistics for each ten-day period at each computational point. However, this work still remains unfinished and the whole thing is being restudied by WARPO.

In recent times, modelling has expanded considerably to cover groundwater, flood management, flood forecasting and warning, urban drainage, and coastal dynamics. WARPO is responsible for monitoring the implementation of NWMP and its evaluation. This will involve assessing impact of different activities against their set targets. For carrying out this responsibility, WARPO would need to develop a number of models that are under process.

To meet the needs of IWRM, mathematical modelling needs to gain more spatial coverage and temporal depth. The models developed in Bangladesh are flood oriented and are unable to yield satisfactory results for dry season flow characteristics. Low flow models need to complement the flood models. As far as spatial coverage of models is concerned, the time has come to think in terms of river basins. The existing river model covers the river system within Bangladesh that does not account for the entire stream flow entering the country from the upstream. It is an imperative to extend the model to cover the entire GBM basin.

Blue Accounting: Another approach for assessment of water resources for integrated management was initiated by a Dutch-assisted project and is now being pursued by its successor, the Centre for Environmental and Geographic Information Services (CEGIS). Designated as blue accounting, this approach takes into account the multiple functions of the water resources system (WRS) of a country and their long-term performance. This then provides the framework for assessing the impact of proposed interventions in terms of national objectives. Such interventions can be structural, affecting the conditions of the WRS or non-structural, focused on the use of the WRS. The approach focuses on a function value system and its successful implementation will add to the inventory of tools for WRS assessment (EGIS–II, 2001).

Environmental Impact Assessment: Over the past few years, awareness about environmental protection has been gaining ground among the different stakeholders involved in the water sector. The Bangladesh government has set-up the Department of Environment (DOE) as the primary institution for environmental management. The Environmental Conservation Act of 1995 designates the DOE as the responsible body for enforcing the environmental impact assessment (EIA) procedures. These procedures are outlined in the EIA Guidelines for Industries of 1997. The title of the guidelines are quite misleading in as much as these cover a good number of water sector interventions like different kinds of water control structures, water supply and sewerage treatment, and roads and bridges. The impacts of these interventions are considered quite heavy on the environment and the most stringent EIA processes have been prescribed for them.

Alongside the DOE guidelines, another set was prepared under the flood action plan for water sector projects that was duly approved by the DOE and its controlling ministry. In addition, an EIA manual was also drafted to assist people concerned with the project who were unfamiliar with the subject. All of these have created some confusion as to which procedure to follow. To remove these difficulties, WARPO and the DOE are trying to work out a new set of guidelines for the water sector. Under the SEMP, the DOE has also started drafting eighteen sets of sectoral EIA Guidelines.

The burning issue in the matter of EIAs is not so much the formulation of new laws and regulations but implementation of the existing ones. The DOE has remained under-resourced for the past couple of years and does not have the institutional capacity to carry out its mandate. Whatever environmental management is taking place in the country is largely driven by donors who impose their own guidelines on the implementing agencies. This is hampering the process of institutionalization. However, a number of institutional strengthening processes are under way and the situation is expected to improve on completion of these activities in couple of year's time.

MANAGEMENT PROCEDURES

Under this cluster would be discussed matters relating to prioritization, guidelines, and economic instruments.

Prioritization Issues: The NWMP has provided a sound basis for short, medium, and long-term perspective plan for integrated management of water resources. The plan covers the activities of thirty-five agencies with a proposed outlay of US$ 18.75 billion over the next twenty-five years. It has provided options for meeting needs of individual components at a micro level as well as options for inter-sectoral and intra-sectoral components at the macro level. These various options have time and cost implications. Activities under the plan will have to be adjusted to accommodate political preferences in an ever-changing political environment and to budgetary constraints which is also a recurrent contingency in Bangladesh. Under these uncertainties, in each plan period, there has to be a core programme that will be implemented at any cost to meet the objectives of IWRM. This brings in the question of prioritization whereby some activities are selected for immediate action leaving others behind for consideration in the next round. Determining priority is intended to be a highly rational act: other things being equal, only those projects are to be selected that come closest to meeting the national goals and demonstrate the prospect of delivering the maximum benefits vis-á-vis the other contenders.

Planning has suffered hugely in Bangladesh due to lack of a political will to enforce strict planning and fiscal discipline. Every successive regime has taken recourse to over-programming, mainly to accommodate rather belated political requests. These late additions are included in the planning document without subjecting them to the rigour of a proper appraisal (Mahmud, 2002). The irony is that sometimes these unplanned and unappraised projects get priority in fund allocation over the regularly programmed ones. These ad hoc decisions put the planning process under severe stress, cause implementation delays and cost overruns, and bring about unwanted distortions in the economy. Rationalization of public expenditure is a major issue in the financial management of the economy of Bangladesh. The implementation of the NWMP will have to find suitable mechanisms to either contain or contend with the issue.

Guidelines: Since IWRM seeks to reorient the process of planning and implementation, these activities need to be carefully guided to keep them on course. What are loosely termed as 'guidelines' are nothing but a systematic presentation of a set of instructions for carrying out a particular segment of activity in the project cycle. Guidelines are intended to ensure that policy objectives are incorporated in project plans and designs and are subsequently followed during project implementation.

The NWP has mentioned the following guidelines for project planning and design:

1. *Guidelines for Project Assessment (GPA).* This is required for the planning and feasibility study of water sector projects to ensure that the projects are consistent with the national and sector objectives, particularly relating to poverty alleviation.
2. *Guidelines for People's Participation (GPP).* This is to ensure people's participation at all stages of project preparation and implementation.
3. *Guidelines for Environmental Impact Assessment (EIA).* This is to ensure that no harm is done to the natural environment by interventions in the water regime and to suggest mitigation measures, if some damage is unavoidable in greater public interest.
4. *Guidelines for Participatory Water Management (GPWM).* The aim is to ensure stakeholder participation in the planning, implementation, and management of water resources projects.

Fortunately, all the above guidelines are in force in the country for water sector project management. Since water management precepts have undergone radical changes in the past few years, the corresponding changes in practice will need a thorough reorientation of the concerned officials for these to come to full force.

Economic Instruments: Water has two types of uses of a fundamental nature: sustaining the ecosystem and servicing mankind as a natural resource having multiple uses. Though water is a renewable resource, it is scarce and it must be used efficiently and in a sustainable manner. The use of economic instruments in water management aims at ensuring its efficient use. The economic tools that are best applied in the water sector relate to the 'user pays' and the 'polluter pays' principles. Both these policies have been endorsed by the NWP.

The 'user pays' principle has been in practice for water supply services in all major cities and secondary towns. The current tariff structure covers only a part of the cost. Tariffs that provide for full cost recovery could be phased in over a number of years. The rural water supply does not levy any charges but it may be necessary to do so in the immediate future for sustainability. However, in the case of large surface water irrigation projects, the 'user pays' principle has been the least effective despite the practice of levying water rate for the past half a century. The willingness to pay is bound up with questions of quality of service and the control of the beneficiaries in the use of funds so collected. Resolution of these two problems may improve the situation in future.

For application of the 'polluter pays' principle, appropriate policy and legal framework are in place. The critical issue for strict enforcement is building capacity of the DOE. A number of initiatives are under process and implementation can be expected shortly.

INSTITUTIONAL ISSUES

Institutions, so far as these relate to water resources, may be defined in terms of both the basic set of rules comprising laws, regulations, and customs as well as organizational arrangements (Guy Le Moigne et al., 1994). The legal issues have already been discussed and this sub-section will cover the organizational part. This will be done by brief discussions on level of action, management boundaries, and capacity building.

LEVEL OF ACTION

Bangladesh is administered by a unitary form of government and it functions through different tiers of administration at regional, district, sub-district, union and, in a few cases, village levels. There are government procedures that clearly delineate the duties and responsibilities of officials at each jurisdictional level. If one goes by the book, government business at the centre is conducted through a number of ministries who, in turn, are assisted by agencies under their control. The core functions of the ministries consist of policy making, and supervision and monitoring of policy implementation. The head offices of the agencies provide executive direction in the implementation of policies to the field offices that are responsible for the actual delivery of service. It is at this level that the government comes in direct contact with members of the public.

In reality, the different administrative layers hardly operate in the manner prescribed in the books. Had they done so, many of the problems now being faced by the water sector would not have arisen at all. Administration in Bangladesh is dominated by strong centralization, to the extent that even routine matters are not disposed off in the field offices. Similarly, there is hardly any feedback from the field offices to their headquarters on different programmes and activities under implementation. It is a one-way top-down flow of information: everything is orchestrated according to the directives received from above. This tendency has not only hampered capacity building of bureaucracy working at local level but has also effectively denied active participation of the stakeholders for better sustainability of projects.

Despite these institutional problems, both the NWP and the NWMP have preferred working with the existing government structure as a long-term strategy for institutional development in Bangladesh. However, for better coordination at the centre and decentralized delivery system at the project level, new organizational arrangements are proposed. On the recommendations of the NWP, the National Water Resources Council (NWRC) has been reconstituted with a fresh mandate that provides policy coordination on all matters relating to water management. It mainly operates through its executive arm, the executive committee of the NWRC (ECNWRC) consisting of a number of Cabinet ministers in charge of water-related ministries. Secretarial and technical support to both the ECNWRC and the NWRC are provided by the WARPO. These new arrangements are already in place and are working satisfactorily.

There is less success with the intended reforms at the field level. The NWP has enunciated two basic principles for ensuring a decentralized delivery system. The first one calls for separation of policy, planning, and regulatory functions from implementation and operational functions at each level of government. The other one envisages field-level activities to be carried out by local government institutions, community-based organizations, and the private sector. The first principle is theoretically in force at the central government level and its abuse can only be minimized by a committed political and bureaucratic leadership willing to support a reorientation of the bureaucratic processes in the country. At the agency level, this has been partly implemented. Independent boards of directors have been constituted and are functioning in the two principal water sector agencies, the Bangladesh Water Development Board and the Dhaka

Water and Sewerage Authority. There are many others who need to review their corporate structure for greater accountability and transparency. The second principle aims at a decentralized delivery system. For water resources management, decentralization has three dimensions. At one level, it means transferring activities that are of such a scale and impact that these can be properly managed by the local government institutions. At another level, it means either deconcentration of centralized offices or delegation of power from a central authority to its subordinate offices in the field. At yet another level, size, cost, and technology considerations will dictate retaining some activities with the centre, allowing beneficiary organizations or private companies to manage them either by themselves or in conjunction with the concerned water sector institutions. These propositions are still being worked out and have not yet been tested in the field.

MANAGEMENT BOUNDARIES

The concept of management boundaries has two dimensions—one relates to the respective functional jurisdictions of the public sector agencies and the other to a determination of the domains between the public sector and the other kinds of organizations.

Government functions are allocated among the different ministries under an 'allocation of business'. The ministries retain the core functions as defined in the rules of business and delegate the remaining ones to the different agencies under its control. All of these are intended to clearly define the management boundaries of the concerned agencies. Unfortunately, since its last review in 1962, there has not been another comprehensive review of this important document. Meanwhile, the nature and scope of public administration have been changing so fast that emerging new activities have remained unallocated by default. This engenders a lot of controversy on jurisdictional matters. It is interesting that in cases where some material benefit is involved, there are many claimants to jurisdiction. When it comes to delivering a service, agencies are prone to disclaim any responsibility (Huda, 2001). Agencies under the different ministries are not only habituated to infringe into the domains of other agencies, they also work unilaterally on highly interrelated matters. Unilateral actions based on departmental priorities have already done lot of damage to the ecosystem of the country. For streamlining the jurisdictional matters, a review of the 'allocation of business' and the corresponding changes in the mandates of the different agencies should be a priority concern of the government. The interministerial steering committees constituted for individual projects could play an important role in conflict resolution. Above all, the ECNWRC is emerging as an important forum for sorting out delicate and contentious interministerial matters.

The NWP has laid down a number of principles for a redefinition of the respective domains of the public and private sector over time. It has pleaded for progressive withdrawal of central government agencies from activities that can be performed better by local institutions and the private sector. It calls for public participation at all stages of the project cycle and handover management of certain types of projects to the beneficiaries. It also advocates contracting out agency functions to private firms who can do it better and at a lesser cost. All these are long-term projections and the process of their institutionalization can only begin when some of these principles are actually put to practice.

CAPACITY BUILDING

IWRM sets in motion a radically different way of doing things. This is accomplished over time by carrying out the necessary institutional and legal reforms to match them to the emerging requirements. The gaps in skills and logistics are identified and steps are taken to strengthen capacity on a continuing basis.

The NWP and the NWMP have set a long and challenging agenda for reforms. Quite a number of these have already been implemented. Notable among these is the promulgation of the BWDB Act of 2000, Guidelines of Participatory Water Management, and the new service rules of the BWDB. Under the new Act, an external policy board has been constituted and there is a programme of transfer of ownership of small projects to the local government and transfer of management of medium-scale projects to the beneficiaries. The Act also stipulates institution of joint management of large projects by the concerned public sector agency and the beneficiaries and, in suitable cases, leasing these out to private parties. The public sector water institutions have never in the past been called upon to handover ownership of their projects to other entities nor do the beneficiaries or the private sector have any experience of managing FCD/FCDI projects. Environmental impact assessments and people's participation at all stages of the project cycle have been made mandatory for processing of water sector operations. These are relatively new rules of engagement with which the concerned parties are not very familiar. No matter how well intentioned the provisions of the law may be, these will be ineffective without a thorough reorientation and skills development of the concerned organizations.

In the context of Bangladesh, institutional capacity has to be built in a number of institutions involved in the reform process by tapping resources from different sources. At the least, these will be the water sector institutions, the community organizations, the NGOs, and the private sector. The specific areas where capacity building activities would concentrate will include human resources development of all the concerned organizations, institutional development and analysis, legislation, water resources planning, data management and interpretation, simulation modelling and other analytical techniques, socio-economic analysis and skills, community skills, and monitoring and evaluation.

In-house training to the officials of most of the water sector institutions is given a very low priority. Most noticeable is the lack of concern in running the newly set-up training institute of the BWDB. The high-quality infrastructures developed for this purpose remain grossly underused. The condition of the other training institutes does not give any better reason for solace. All efforts at reform of the water sector may be negated if effective steps are not taken for capacity building of all stakeholders in the water sector.

Conclusions

The assessment carried out under this study shows tremendous progress made by Bangladesh in creating the enabling environment for launching an IWRM programme. The official declaration of the NWP was a major landmark in trying to bring about order and rationality in water resources management by providing the necessary guidelines. Unlike the fate of many other policy documents, this initiative did not stagnate at pious pronouncements only. A NWMP has been carefully prepared to implement the directives of the NWP through elaborate consultations with all stakeholders. Both the policy and the plan have assigned specific roles to local institutions, community organizations, and the private sector, limiting the role of the central government and its agencies to creating the boundary conditions. Following the guidelines of the NWP, the mandate of the Bangladesh Water Development Board (BWDB), the largest water sector implementing agency, has been thoroughly reviewed and a new BWDB Act, 2000, has been promulgated. This was preceded by a long overdue reorganization of the structure of the institution that rationalized its structure, staffing, and couple of subsidiary rules and regulations. A revision and restructuring of WARPO, another important water sector institution responsible for updating and implementation monitoring of the NWMP, is under process. A *Guidelines for Participatory Water*

Management, prepared by an interministerial task force has been approved by the government and is now being followed in project management. All of these and many other activities now under process, give the hope that Bangladesh now is ready for undertaking an IWRM programme. However, a number of risks and uncertainties still lurk on the horizon. IWRM challenges the fragmented approach to water management by a number of autonomous agencies and requires all concerned agencies to agree on a common programme for implementation in the field. The process of the formulation of NWMP has shown that despite conflicting interests, different agencies and other stakeholders can agree, on a common programme. Integrated planning at the centre has not been problematic. Things may be different when actual implementation starts at the local level. Given the long tradition of departmental execution of projects, it remains an open question as to how much accommodation for an integrated approach will persist at the ground level. There is also the possibility that appropriate organizational arrangements for implementation of integrated projects could remove the causes that discourage inter-agency cooperation. A recent case study has found that couple of projects have successfully been executed *jointly* by a number of agencies under formalized procedures. Under these arrangements, each agency is allowed to execute its own component departmentally while at the same time is required to carry out its activities within the agreed planning and implementation framework (Huda, 2002).

The important thing for Bangladesh is now to start implementing the NWMP. It is impractical to think that all the theoretically derived conditions should be fulfilled prior to initiating IWRM. The best approach is learning by doing. As the programme moves ahead, many unanticipated constraints may appear. The situational context will determine how to deal with those. The risk of that kind of contingencies should not deter our efforts at initiating IWRM. After all, this is a new kind of management approach that seeks to build on experience and knowledge surrounding the complexities of integrated water management.

References

Adhikari, K.D., Q.K. Ahmad., S.K. Malla., B.B. Pradhan, Khalilur Rahman, R. Rangachari, K.B. Sajjadur Rasheed, and B.G. Verghese, *Cooperation on Eastern Himalayan Rivers: Opportunities and Challenges*, New Delhi: Konark Publishers Pvt Ltd, 2000.

Ahmad, Q.K., Asit K. Biswas., R. Ranghachari, and M.M. Sainju (eds), *Ganges-Brahmaputra-Meghna Region: A Framework for Sustainable Development* Dhaka: The University Press Limited, 2001.

Ahmad, Q.K., B.G. Verghese, Ramaswamy R. Iyer, B.B. Pradhan, and S.K. Malla (eds), *Converting Water into Wealth: Regional Co-operation in Harnessing the Eastern Himalayan Rivers* Dhaka: Academic Publishers, 1994.

Ahmed, K.M., 'Arsenic Contamination in Groundwater and a Review of the Situation in Bangladesh' in A.A. Rahman and P. Ravenscroft (eds), *Groundwater Resources of Bangladesh*, Dhaka: The University Press Limited, 2000.

Bangladesh Action Plan for Flood Control, Washington D.C.: World Bank Asia Region Country Department I, 1989.

Bangladesh: Climate Change and Sustainable Development, Dhaka: World Bank, 2000.

Bangladesh Land and Water Resources Sector Study, Washington D.C.: International Bank for Reconstruction and Development, Asia Projects Department, 1972.

Bangladesh-Nepal Joint Study Team *Report on Mitigation Measures and Multipurpose Use of Water Resources*, Dhaka and Kathmandu: Ministry of Irrigation, Water Development and Flood Control, Government of Bangladesh, and Ministry of Water Resources, His Majesty's Government of Nepal, 1989.

Bangladesh: Plan of Action on National Agricultural Policy, BGD/00/006: SPPD Final Report. Ministry of Agriculture, Government of Bangladesh, Food and Agriculture Organization and the United Nations Development Programme.

Bangladesh: The Sustainability of Agricultural Growth, Washington D.C.: World Bank, Agriculture and Natural Resources Division, 1995.

Bangladesh Water and Flood Management Strategy: An Update following the Signing of the Ganges Water Sharing Treaty, Dhaka: Ministry of Water Resources, Government of Bangladesh, 1998.

Bangladesh: Water Sector Review, Manila: Asian Development Bank, 2000.

Biswas, A.K., and J.I. Uitto (eds), *Sustainable Development of the Ganges-Brahmaputra-Meghna Basins*, Tokyo: United Nations University Press, 2001.

Biswas, A.K., and T. Hashimoto (eds), *Asian International Waters: From Ganges-Brahmaputra to Mekong*, Bombay: Oxford University Press, 1996.

Brammer, H., 'Environmental Aspects of Flood Protection in Bangladesh', *Asia-Pacific Journal on Environment and Development*, 2:2, 1995, pp. 30–42.

Brammer, H., 'Flood, Flood Mitigation and Soil Fertility in Bangladesh', *Asia Pacific Journal on Environment and Development*, 2:1, 1995, pp. 13–24.

Brandon, C., 'Environmental Degradation and Agricultural Growth' in Rashid Faruqee (ed.), *Bangladesh Agriculture in the 21st Century*, Dhaka: The University Press Limited, 1998.

Choudhry, Y.A., 'Integrated Water Resources Management in Bangladesh: From Policy to Plan and Execution', in Mustafa Alam and Rob Koudstaal (eds), *Integrated Water Resources Management: Perspectives from Bangladesh and the Netherlands*, Dhaka: The University Press Limited, 2001.

Chowdhury, J.U., Mohammad Rezaur Rahman, and Masfiqus Salehin, *Flood Control in a Floodplain Country: Experiences of Bangladesh*, Dhaka: Institute of Flood Control and Drainage Research, Bangladesh University of Engineering and Technology, 1996.

Coastal Zone Management: An Analysis of Different Policy Documents, Dhaka, Project Development Office, Integrated Coastal Zone Management (PDO-ICZM), 2001.

Datta, A.K., *Planning and Management of Water Resources: Lessons from Two Decades of Early Implementation Project*, Dhaka: The University Press Limited, 1999.

Draft Development Strategy: National Water Management Plan Project, Main Report, Volume 2, Dhaka: Water Resources Planning Organization (WARPO), 2000a.

Draft Development Strategy: National Water Management Plan Project, Annex M, Appendix I, Dhaka: Water Resources Planning Organization (WARPO), 2000b.

Draft Development Strategy: National Water Management Plan Project, Annex C, Dhaka: (WARPO) Water Resources Planning Organization (WARPO), 2000d.

East Pakistan Water and Power Development Authority, *Master Plan*, Volumes I and II, Dhaka: EPWADA, 1964.

Eighteen District Towns Water Supply Project, Final Report, Dhaka: Department of Public Health Engineering, 2000.

Fisheries Specialist Study, *Main Report*, Volume I, Flood Action Plan, Northeast Regional Water Management Project (FAP 6), Dhaka: Flood Plan Coordination Organization (FPCO), 1994.

Groundwater Studies for Arsenic Contamination in Bangladesh, Rapid Investigation Phase, Final Report, Dhaka: Department of Public Health Engineering, 1999.

Guidebook for Integrated Water Resources Management: Concepts and Tools, *Draft Version*, Dhaka: EGIS–II, 1964.

Habib, E., *Management of Fisheries, Coastal Resources and the Coastal Environment in Bangladesh: Legal and Institutional Perspectives*, Manila: International Centre for Living Aquatic Resource Management, 1999.

Huda, S., 'Institutional Design for ICZM: A Bangladesh Case Study', Paper presented at the Coastal Zone Asia-Pacific Conference held at Bangkok, Thailand during 12–16 May 2002.

Huda, S., *Institutional Review of Selected Ministries and Agencies*, Dhaka: Project Development Office, Integrated Coastal Zone Management, 2001.

'Inventory and Rural FCD', in *Draft Development Strategy: National Water Management Plan Project*, Annex M., Appendix 2, Dhaka: Water Resources Planning Organization (WARPO), 2000e.

Integrated Water Resources Management. TAC Background Papers No 4, Technical Advisory Committee (TAC), Global Water Partnership, 2000.

Khan, A.A., 'Freshwater Wetlands in Bangladesh: Opportunities and Options' in A. Nishat Z. Hussain, M.K. Roy, and A. Karim (eds), *Freshwater Wetlands in Bangladesh: Issues and Approaches for Management*, Gland, Switzerland: International Union for Conservation of Nature and Natural Resources, 1993.

'Land Use Conflicts arising from Shrimp and Prawn Cultivation', in *Draft Development Strategy: National Water Management Plan Project*, Annex F, Appendix I, Dhaka: Water Resources Planning Organization (WARPO), 2000c.

Le Moigne, G., Ashok Subramanian, Mei Xie, and Sandra Giltner (eds), *A Guide to the Formulation of Water Resources Strategy*: Washington D.C.: World Bank, 1994.

Mahmud, W., 'State, Market and Institutional Transformation', Paper read at the annual conference of the Bangladesh Economic Association held at Dhaka during 20–22 September, 2002.

Major Strategies and Programme Framework for Agricultural Development in Bangladesh, Dhaka: Ministry of Agriculture, Government of Bangladesh, 1997.

McIntire, J., 'The Prospects for Agricultural Growth', in Rashid Faruqee (ed.), *Bangladesh Agriculture in the 21st Century*, Dhaka: The University Press Limited, 1998.

National Agriculture Policy, Dhaka: Ministry of Agriculture, 1999.

National Water Management Plan: Main Report, Volume 2, Dhaka: Water Resources Planning Organization, 2001.

National Water Plan, Phase I, Dhaka: Master Plan Organization, 1987.

National Water Plan, Phase II, Dhaka: Master Plan Organization, 1991.

National Water Policy, Dhaka: Ministry of Water Resources, Government of Bangladesh, 1999.

Options for the Ganges Dependent Area, Volume I, Main Report, Dhaka: Water Resources Planning Organization (WARPO), 2002.

Pagiola, S., *Environmental and Natural Resource Degradation in Intensive Agriculture in Bangladesh*, Washington D.C.: World Bank, 1999.

Paul, R., 'Gender and Water Supply, Sanitation and Environment', Bangladesh Water Partnership Workshop on Gender and Water held at Dhaka on 14 November 1999.

Rahman, M.R., 'Towards Sustainable Development: Land Resources in Bangladesh', A Background Paper prepared for National Conservation Strategy, Dhaka: International Union for Conservation of Nature and Natural Resources, 1990.

Rasheed, S., 'Overview of Water Resources Management and Development in Bangladesh', in *Seminar Proceedings of International Seminar on Water Resources*

Management and Development in Bangladesh with particular reference to the Ganges River held at Dhaka, 8–10 March 1998, Dhaka: Ministry of Water Resources, Government of Bangladesh, 1998.

Rogers, P., 'The Economic and Financial Context', in Arriens, Wouter Lincklaen, Jeremy Bird, Jeremy Berkoff, and Paul Mosley (eds), *Towards Effective Water Policy in the Asian and Pacific Region*, Volume III, Manila: Asian Development Bank, 1996.

Siddiqi, M.H., 'Options for Development in the Ganges Dependent Areas of Bangladesh—Historical Perspectives', in *Seminar Proceedings of International Seminar on Water Resources Management and Development in Bangladesh with particular reference to the Ganges River held at Dhaka, 8–10 March 1998*, Dhaka: Ministry of Water Resources, Government of Bangladesh, 1998.

Soussan, J., A. Datta, and P. Wattage, *Community Partnership for Sustainable Water Management: Experience of the BWDB Systems Rehabilitation Project*, Six Volumes, Dhaka: The University Press Limited, 1998.

Towards Sustainable Development: The Bangladesh National Conservation Strategy: Dhaka Government of Bangladesh and the IUCN, 1997.

Water Resources Management in Bangladesh: Steps Towards a New National Water Plan, Dhaka: World Bank, Rural Development Sector Unit, South Asia Region, 1998.

Water Resources Management, A World Bank Policy Paper, Washington D.C.: World Bank, 1993.

Wilde, K. de, *Out of the Periphery: Development of Coastal Chars in Southeastern Bangladesh*, Dhaka: The University Press Limited, 2000.

5

STATUS OF INTEGRATED WATER RESOURCES MANAGEMENT IN NEPAL: AN OVERVIEW

Iswer Raj Onta

Introduction

The physical existence of innumerable rivers, estimated at some 6000 in number and major perennial rivers emanating from the lofty Himalayas has prompted many in Nepal to label the nation as abundant in water resources. But to convert this abundance to an useful natural resource, Nepal must be able to harness the rivers for the benefit of not only all citizens of the country but develop it so as to maintain the entire ecological chain on which all life and life-sustaining systems depend. So far, this process of converting the river water to resources has not been encouraging. More than 25 per cent of Nepal's population do not have access to safe drinking water; more than 50 per cent do not have sanitation coverage; over 50 per cent of cultivated areas has no irrigation facilities and around 80 per cent of the population are yet to receive electricity.

However, river water is considered the only natural resource that Nepal can depend on and is regarded as one of the key strategic resources which has the potential to be the catalyst for all-round development and economic growth of the country, thereby reducing the overarching problem of poverty the nation is facing at present (it is estimated that over 38 per cent of the country's population lived below the nationally defined poverty line in 1999).

The major river systems of Nepal viz. Karnali, Sapta Gandaki, Saptakosi, and Mahakali account for around 80 per cent of the total run-off of over 220 billion cubic metres (BCM). These major rivers are deeply incised (tributary of Sapta Gandaki, Kali Gandaki, carves one of the deepest gorge in the world between the Himalayan peaks of Dhaulagiri and Annapurna, both more than 8000 metres above the mean sea level). These rivers are difficult to harness for substantial uses unless mammoth projects are designed for multipurpose uses, which, in turn, requires huge investments. The water control structures in such rivers are at a low level. The tourism sector, however, uses these large rivers for limited recreational uses like white water rafting and kayaking. Thèse large rivers are also used for small-time fishing by fishermen who live along the villages by the river high above the high flood level.

The people of Nepal, by and large, depend on tributaries of these large rivers for their water needs, be it for domestic, irrigation, industrial, or other purposes. These rivers are generally not perennial and flows vary tremendously between the wet (monsoon) and dry seasons. The wet monsoon season lasts for a four-month period from June to September.

The other eight months of dry periods are very crucial in terms of water use, and scarcity of water is felt in many locations during this period. Seasonal local scarcity is manifesting itself more frequently. Water disputes surface during this period as well. It is not unusual to read in newspapers that such disputes turn into gun battles between village groups. It is also sometimes reported that people have abandoned their villages and migrated towards areas where water is available, because the bare minimum water for their survival is sometimes not available.

Groundwater, estimated at twelve BCM, is at present available in plenty in most of the flat Terai plains in the southern part of the country. Extraction of groundwater for conjunctive irrigation and for drinking, as well as for industrial purposes is practised as a matter of course in the Terai region. The problem of arsenic contamination in the groundwater has surfaced in recent times.

It has now been recognized at certain levels that freshwater is a finite and vulnerable resource. Nepal's existing population of over twenty-three million is expected to double in the next thirty years and the demands for water are continually increasing for different purposes. Demand for drinking water is increasing at a much faster rate as the rate of urbanization is increasing at an accelerated pace and urban population tends to demand relatively more water than rural population.

At the same time, the river pollution in and around urban centres too is increasing at a much faster rate and people are demanding a clean river environment. Urban poor, slum dwellers, and squatters are demanding attention towards their water requirement, which they consider as their just right.

The increase in population is also putting pressure on other natural resources including forests. Deforestation has adversely affected many natural watersheds, resulting in degradation of water sources including spring sources in the hills and mountains, on which many rural people are dependent upon.

Heavy rainfalls during monsoon cause wide variations in river flows, which tend to induce land erosion and landslides. The frequency of devastating landslides, which sweep out entire villages in hills and mountains, is increasing each year with loss of human life and livestocks. Frequency of damaging floods are increasing too.

All such events and situations have forced the government to come up with the Water Resources Sector Strategy encompassing the development and management of water resources in a holistic and systematic approach, relying on the integrated water resource management system (Figure 5.1).

Overview of Nepal's Water Resources

The Himalayan kingdom of Nepal has a total land area of 147,480 km^2 with diverse topographical characteristics, climate, geology, and land use system. A predominantly mountainous country, it is criss-crossed by innumerable rivers. The total number of rivers as mentioned earlier is estimated at 6000, and the total length of water courses is about 45,000 km.

Nepal experiences the monsoonal climate. The average annual precipitation is about 1500 mm (80 per cent of which occurs during the monsoon season). The spatial and temporal nature of rainfall results in large variation in surface water flows, which, in turn, creates surpluses at certain times and places, primarily during the four monsoon months and shortages at others. Besides surface water, Nepal is also endowed with extensive groundwater resources, found in most of the Terai plain and in some mid-hill valleys like the capital city of Kathmandu. The total dynamic groundwater reserve is estimated at 8.8 BCM. In the Terai, present annual withdrawal of groundwater, for various purposes, is estimated at a mere one BCM. In Kathmandu Valley, in contrast, the total annual groundwater abstraction is presently estimated at around twenty-three million cubic metres (MCM), whereas the recharge is estimated by some experts, at about five MCM. In spite of such extraction, Kathmandu is facing a shortage of drinking water supply and an inter-basin water transfer project from Koshi basin, called Melamchi water supply project is being planned and implemented.

In terms of per capita availability, average annual per capita availability of renewable water resources for the present Nepalese population comes to about 9400 m^3, which is above the global average of about

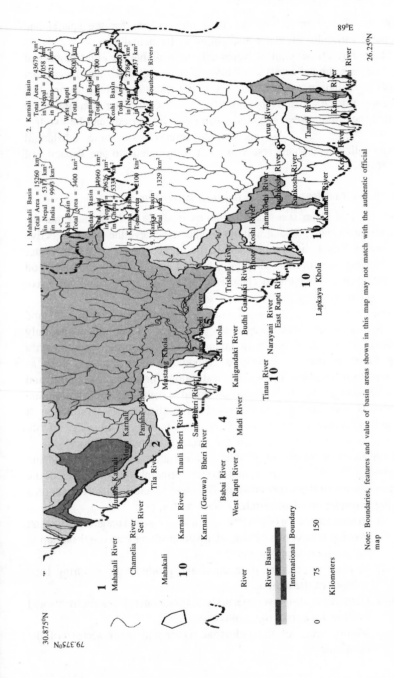

Figure 5.1: Major river basins of Nepal

Note: Boundaries, features and value of basin areas shown in this map may not match with the authentic official map

6600 m^3. This figure, however, does not account for the spatial and temporal variability in water availability or for actual uses.

Since only 20 per cent of water will be available for eight months of any one year, the per capita availability drops to 1800 m^3. This means that Nepal's population will remain close to the water stress level during this critical period. As such, Nepal must find ways to store the monsoon water so that the water stress situation during the dry periods can be averted. So far, Nepal has only one storage reservoir of some substantial nature called Kulekhani reservoir with a designed storage capacity of 83 MCM^3.

Water Sector Strategy and Enabling Environment

Realizing that water is a strategic natural resource of Nepal and that the business-as-usual approach will not be able to achieve the stated objective that 'every Nepali citizen, now and in the future, has access to safe water for drinking, appropriate sanitation, and enough water to produce food and energy at reasonable cost', Nepal embarked upon its first comprehensive water sector strategy.

The long and hard process of strategy formulation began in 1995 and culminated in the 'Water Resources Strategies, Nepal' document duly approved by the government in January 2002. The process took such a long time because the principles embodied in the strategy, such as establishment and development of 'institution' for strategic planning and that the strategy was to be based on broad stakeholder consultation and consensus, was satisfactorily exercised during the strategy formulation. It is said that about two thousand (Nepalese) people of all walks of life participated in various stages of the strategy formulation process.

Prior to the development of water sector strategy, Nepal developed and adopted a set of objectives and policy principles that provided the framework for strategy formulation. The eight specific objectives adopted for the formulation of the strategy were as follows:

  (i) Reduce incidence of poverty, unemployment, and underemployment;
 (ii) Provide access to safe and adequate drinking water and sanitation for ensuring health security;
(iii) Increase agriculture production and productivity, ensuring food security of the nation;
 (iv) Generate hydropower to satisfy national energy requirements and to allow for export of surplus energy;
  (v) Supply needs of the industrial sector and other sectors of the economics;

(vi) Facilitate water transport, particularly connection to a seaport;

(vii) Protect the environment and conserve the biodiversity of natural habitat; and

(viii) Prevent and mitigate water-induced disasters.

Similarly, the following eight policy principles were adopted to develop the water strategy:

(i) Development and management of water resources to be undertaken in a holistic and systematic manner, relying on integrated water resources management (IWRM);

(ii) Water utilization to be sustainable to ensure conservation of the resource and protection of the environment. Each river basin to be managed holistically;

(iii) Delivery of water services to be decentralized in a manner that involves autonomous and accountable agencies (e.g., public, private, community, and user-based agencies);

(iv) Economic efficiency and social equity to guide water resources development and management;

(v) Participation of and consultation with all stakeholders to constitute the basis of water sector development;

(vi) Sharing of water resource benefits among the riparian countries to be on an equitable basis for mutual benefits;

(vii) Institutional and legal frameworks for coordination and transparency to be an essential feature of water sector management; and

(viii) Wider adoption of the best existing technologies and practices, and rapid innovation and adoption of both institutional arrangements and new technologies, to be ensured.

Based on these set of objectives and policy principles, the water sector strategy was formulated using participatory approach. This approach involved formulating a strategic goal and purposes to contribute to that goal. To achieve the purposes, a set of ten strategic outputs were identified.

The strategic goal has been defined as 'Living conditions of Nepali people are significantly improved in a sustainable manner'. The purposes, which will contribute to the goal was formulated such that some breakthrough, like achievements of 'basic needs' in all sectors, needed to be achieved faster to be able to maintain the support of all stakeholders. For such breakthrough to occur, a short-term (five years) purpose and a medium-term (fifteen years) purpose are defined and corresponding outputs and activities are also identified. The strategy formulation has sought to cover a period of thirty years. The long-term (twenty-five years) purposes have been defined as 'benefits from water resources are maximized in Nepal in a sustainable manner'.

The water resources strategy has defined the following ten strategic outputs:

(i) Effective measures to manage and mitigate water-induced disasters are functional;

(ii) Sustainable management of watershed and aquatic ecosystems achieved;

(iii) Adequate supply of and access to potable water and sanitation and hygiene awareness provided;

(iv) Appropriate and efficient irrigation available to support optimal, sustainable use of irrigable land;

(v) Cost-effective hydropower developed in a sustainable manner;

(vi) Economic uses of water by industries and waterbodies by tourism, fishery, and navigation optimized;

(vii) Enhanced water-related information systems are functional;

(viii) Appropriate legal frameworks are functional;

(ix) Regional cooperation for substantial mutual benefit achieved; and

(x) Appropriate institutional mechanism for water sector management is functional.

For better understanding by the planners and stakeholders the strategic outputs have been categorized into three functional aspects. The first two outputs are categorized under 'security', i.e., security from water and security of water; the next four outputs are categorized under 'use', i.e., drinking water supply and sanitation, irrigation, hydropower, and other uses; and the last four outputs are categorized under 'mechanism', i.e., information systems, legal frameworks, regional cooperation, and institutional support.

Each of these 'outputs' is achieved only when a set of activities encompassing physical, managerial, economic, and institutional aspects are successfully implemented. The strategy has identified a total of sixty-four major activities. The strategy has also enumerated the preconditions, assumptions and risk associated with it. The indicators to measure the effectiveness of the activities have been identified and associated investments have been estimated.

The next phase of the water sector strategy, which has just recently begun, is the preparation of a national water plan based on those identified major activities which are expected to provide guidance towards the implementation of the water resource strategy. The national water plan will also include an environmental management plan for the water sector.

Legal Framework

The National Code of 1853 was the first comprehensive statutory law in Nepal which included the rights of people on the usage of water. Subsequently, some sector-specific acts and regulations came to existence. But it was only in 1992 that a legislation called Water Resources Act came into existence. After that, series of acts and regulations appeared on the scene. Some of these are Electricity Act (1992), Electricity Tariff Fixation Regulation (1993), Water Resources Regulation (1993), Irrigation Regulation (2000), Electricity Regulation (1993), Industrial Enterprises Act (1992), Environment Protection Act (1996), Drinking Water Regulation (1998), and recently the Local Self-governance Act (1999).

Nepal has formulated sub-sectoral policies such as Irrigation Policy (1992 and amended in 1997 and 2003), Hydropower Development Policy (1992, amended in 2001), and Water Supply Sector Policy (1998). Unfortunately, these sub-sectoral policy documents do not profess the principle of IWRM in the present form. However, Nepal has not yet formulated a comprehensive policy for the water sector. Due to lack of a comprehensive water policy, many water-related sub-sectoral programmes with inherent biases have not achieved the desired results and, in some cases, programmes have been either entirely stopped or derailed. In the process, the people at large have suffered due to non-delivery or ineffective delivery of water-related services.

Formulation of a comprehensive water policy is one of the urgent requirements identified in the WRS 2002. It is reported that a comprehensive water policy and harmonization of different policies, acts, and regulations are being currently undertaken under the process for the formulation of the National Water Plan.

The process of integration of sectoral policies is difficult in the Nepalese scenario because the bureaucracy inherits the 'quick fix' and 'fatalistic' culture of the society. The integration of policies and coordinated actions thereof are not only desirable and necessary for the upliftment of society but is a complex and time-enduring task. There seems to be some realization of this in some quarters, but achieving an effective coordination and integration of all sectoral policy frameworks in Nepal is still a long way off.

Some Salient Features in the Acts and Regulations

The Water Resources Act (1992) states that the ownership of all water resources lies with the State and people will have water use rights so that the resource is utilized for creating wealth. The Act also provides the priority order of uses, which are as follows:

  (i)   Drinking and domestic;
  (ii) Irrigation;
  (iii) Agriculture (animal husbandry);
  (iv) Hydropower;
  (v)   Cottage industry, industrial enterprises, and mining;
  (vi) Navigation;
  (vii) Recreational; and
  (viii) Others.

The Act prescribes that a government licence is required for the development of water resources by any agency(ies). However, it is also mentioned that the development of water sources, for individual and collective use by community, for drinking water purposes does not require licences.

The Water Resources Regulation (1993) has identified the District Water Resources Committee (DWRC) as an agency where all water users have to register and licence to use is to be procured after paying a fee. DWRC is headed by the chief district officer (CDO) and comprises of representatives of district-level government offices of agriculture, forest, drinking water, and irrigation; representative from department of electricity development (where appropriate); local government representative from district development committee (DDC), and the local development officer, who also act as member secretary of the committee. DWRC is also to be responsible for monitoring the water resources use and development in each district. The spirit of this provision is to allocate water resources for different users in an equitable and justifiable manner. It could be the precursor of local-level IWRM institution. However, DWRC has not been functional and, in many cases, they are not consulted or just ignored by stakeholders when conflicts arise because the stakeholders' traditional mechanism to resolve issues such as the water rights and uses seems to be working fairly, in many cases, at present. Surprisingly, even a parastatal institution like Nepal Water Supply Corporation (NWSC) does not recognize that it too has to register its use of water with the DWRC. On the other hand, the DWRC in some districts has not been properly set-up, managed, trained, and supported in the concepts of IWRM and water policy principles and, therefore, does not function as planned.

The Act provides mechanism for dispute resolution by constituting committees which are named as 'Water Resources Utilization Investigation Committee' at the national level, and 'Water Source Dispute Resolution Committee' at the district level. Very few stakeholders have knowledge of these statutory arrangements.

One serious concern hindering the application of water management and regulatory mechanism, are the conflicting procedural legislations in various acts and regulations. The Drinking Water Regulation (1998), for instance, has prescribed separate and contradictory procedures with regard to licensing, formation of users group, dispute settlement mechanism, etc. Likewise, the Irrigation Regulation (2000) has also prescribed a separate set of rules and procedures regarding formation and registration of user's association. The authority of the 'licensing' is entrusted to the district irrigation office (DIO), whereas the Local Self-governance Act (1999) has given similar authority to the village development committees (VDCs), municipalities, and DDCs. Similarly, licensing authority to develop hydropower is vested with the Ministry of Water Resources (MOWR) as per Electricity Act but the need to register it with DWRC is not clearly spelled out. These problems have been identified and hence the harmonization of legislations is one of the key activities of the water sector strategy.

Concept of Integrated Water Resources Management

IWRM is a concept for ensuring the integration in the multifunctional use of both surface and groundwater in river basins for the present as well as future generations. IWRM is the integrating tool that can lead the users from fragmented sub-sectoral to holistic cross-sectoral water resource management. IWRM is much broader than traditional water management and includes significant parts of land use planning, agricultural components, erosion control, environmental management, and other policy areas.

IWRM, which in some particular cases, is synonymous to river basin management, is primarily based on hydrological boundaries. It is cross-sectoral in nature and aims at a broad inclusion of stakeholders. It is a process that promotes coordinated development and management of water, land, and related resources. Its goal is to maximize economic and social well-being in an equitable manner without compromising the sustainability of vital ecosystems.

As population grows, the water stresses will be increasingly felt in all sectors, be it drinking water, irrigation, urban, or river environment. Competition for the use of water between different sectors will grow and IWRM is considered a viable means of resolving the conflicting issues while ensuring increased economic productivity, equity, and sustainability of available water resources.

However, it is increasingly evident that operationalizing IWRM is quite difficult, be it in sectoral planning level or at community and village levels.

Some attempts made in Nepal, to understand and operationlize IWRM, will be presented in the following section.

Case Studies

Attempts are made here to document findings as well as processes adopted in four different studies related to IWRM and try to answer questions like how IWRM is defined in case-specific context, what specific aspects are being integrated, by whom and through what process, and what were the expected results. Similarly, the studies have tried to learn lessons from existing systems and find answers to questions like how IWRM improves the management process; what are the institutional arrangements; what were the social, economic, and environmental impacts in the affected river basins and regions.

Nepal is experiencing different models of application of integrated development at the local level, which show that not one particular model will be applicable in all cases. There are many more cases of local-level attempts to integrated development activities which include water-related activities, without consciously realizing that the processes adopted are leading towards IWRM practices.

The case studies mentioned below may not answer all these questions but it is believed that some case studies will throw some interesting insight towards operationalization of IWRM in the context of water management in Nepal.

Indrawati River Basin[1]

Physical Characteristics of the Basin

The Indrawati river basin is a part of the large Koshi river basin and is located in the central region contiguous to Bagmati river basin, wherein lies the capital city of Kathmandu (Figure 5.2). The basin lies in the three administrative districts of Sindhupalchowk, Kabhrepalanchowk, and Kathmandu. The catchment area of the basin is 1240 km^2. The basin has

---

[1] In order to assist the Water and Energy Commission Secretariat (WECS) in its ongoing project activities related to IWRM, IWMI, with the fund provided by Ford Foundation, embarked upon four focussed studies in Indrawati river basin, on areas like 'Formal and Informal Water Institutions', 'Water Accounting Status', 'Social Exclusion and Inclusions', and 'Process Documentation Research (PDR) of the Melamchi Water-Project', with an aim to analyse constraints and opportunities of managing water resources according to the principle of IWRM.

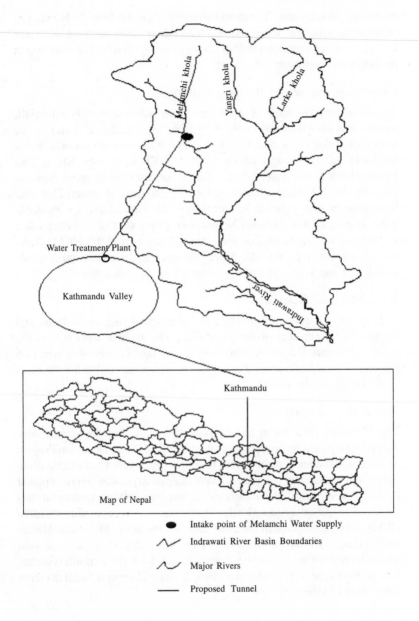

Figure 5.2: Indrawati sub-basin

40 per cent forested area. The Indrawati River originates from the Himalayan region (5863m above mean sea level) and drains into the Sunkoshi river at a place called Dolalghat (626m above mean sea level). The total length of Indrawati river course is 59 km.

HYDROLOGY AND WATER BALANCE SITUATION

The basin receives most of its waters from seasonal monsoon rainfall, which turns into snowfall in the high Himalayas, and contributes to the river flow during hot summer months. Major tributaries of Indrawati River are Larke Khola[2], Yangri Khola, Melamchi Khola, Jyangri Khola, Cha Khola, Handi Khola, and Mahadev Khola. Annual rainfall varies between 3100 mm to 2300 mm with average rainfall estimated at around 2700 mm in an average year in the Indrawati river basin. Annual average available water is estimated at 3373 MCM and only 23 per cent of available water in the basin is consumed. The remaining 77 per cent (2360 MCM) flows out of the basin as run-off. Since the utilizable outflow takes place throughout the year, the basin is denoted as an 'open basin'.

SOCIO-ECONOMIC FEATURES

Agriculture is the main occupation of the people living in the basin and it employs 96 per cent of the population. The average farm size in the basin is less than 0.5 ha. Around 2000 ha. of land is cultivable area (1.6 per cent of total basin area). Paddy, wheat, maize, and millet are the main crops grown in the basin.

WATER USE ACTIVITY

The Indrawati river basin has recently been targetted for several new water development projects, notably the Melamchi Water Diversion Project, an inter-basin water transfer project to supply drinking water to Kathmandu valley, and the recently completed 7 MW (megawatt) Indrawati Hydropower Project. It was estimated that there exist more than 300 functioning farmer-managed irrigation system (FMIS) in the basin. A recent study (WECS/ IWMI, 2000) conducted in the three sample tributaries (Melamchi, Handi, and Mahadev Khola) of Indrawati river shows that in these three sub-basins, there exist nineteen FMIS, twenty-two traditional watermills (Ghatta), four turbine-type watermills, one micro-hydel and several small drinking water supply schemes.

---

[2] *'Khola'* is a generic term, which denotes small rivers.

As population grows, the need for bringing more land under irrigation to sustain the food demands will grow, and additional water use activities like watermills, ghatta, and drinking water schemes will be taken up. The demands for more water will put pressure on the allocation and reallocation of available water resource in the Indrawati river and its tributaries.

MAJOR STAKEHOLDER AND WATER RIGHTS

Irrigation users, ghatta owners, watermill owners, and micro-hydropower users are the major stakeholders. Water user associations and micro-hydro committees are the organized stakeholders who represent their respective members. DDC and VDC are local government institutional stakeholders. NGOs are appearing to be the new generation of stakeholders, who are representing poor and marginalized communities.

As per the local custom, existing water users have the first rights to use the water over the newcomers or latecomers. After drinking water, irrigation gets priority over other uses, which is followed by Ghatta, watermill and/or hydropower. Water rights in Indrawati river basin are secured in three ways:

- Customary rule, first user get first priority (appropriation rights);
- Priority to upstream water users over others (riparian rights); and
- By registration at VDC, or at DDC.

INCLUSIONS AND EXCLUSION IN WATER USE SERVICES

Participation of all stakeholders during negotiations prior to any new investment in water resources project is crucial from the point of view of inclusions or exclusions. Economically poor and marginal stakeholders are generally ignored and excluded from future water use services because proper representation of this section of population during consultation/ negotiation meetings is still not adequate.

In irrigation schemes, exclusion occurs, sometimes, due to geotechnical problems. Siting of intake and canal alignment in the hilly and mountainous topography has to be technically and geologically feasible, and therefore, in some cases, some cultivable lands could be left out of the command area due to technical reasons.

Poor and marginal households are sometimes left out of the service of electricity generated by the community as well as privately owned micro-hydropower schemes because of their inability to pay the monthly minimum charges (about NRs. 25.00 per month). Economic factors seem to be the exclusion criteria in this case.

For drinking water services too, economic factors seem to govern the exclusionary condition. Scattered settlement patterns in the hills are some of the reasons for not getting access to piped drinking water supply.

Ghatta, the traditional watermill, is generally owned and operated by poor households. They are not usually consulted or their voices are not heard properly even when their Ghatta operations are affected due to new water use activities, particularly during some low flow periods. The role of VDC and DDC as well as NGOs then becomes crucial in addressing such issues which concerns the equitable distribution of water resources. The resentments of the excluded communities are the underlying factors for some of the conflicts that surface out in different forms.

Water Shortage and Disputes Resolutions

Water shortage is experienced during the low flow period between February and May. It has been reported that cases of water disputes between Ghatta owners and the irrigation farmer are frequent during these dry periods. However, most of the reported water disputes are said to have been resolved based on customary rules and informal arrangements of water allocation. Some of the customary rules adopted to avoid water disputes in the basin are as follows:

- Ghatta owners close their operations during peak irrigation demand.
- Generally, where water turbine mills are located at the downstream end of the system, the mill owners take responsibility to maintain the canal, which benefit their mill and, in turn, farmers get irrigation water without having to pay for Operation & maintenance (O&M) of canal.
- A distance of 200 metres must be maintained between any two water intake locations.

However, the customary practices and informal arrangement to resolve the conflict may not be sufficient in future as more and more water resources development projects are taken up. The roles of VDC, DDC, and DWRC will emerge as crucial, and formal arrangements and rules and regulations will need to be in place, as envisaged by the Local Self-governance Act for sustainable development of water resources in each district.

Water-related Institutions in the Basin

Both formal and informal institutions exist and are functioning in the Indrawati river basin. The informal institutions are functioning effectively at present as the utilization of water resources use is at a low level. These informal institutions function at community level. Formal government institutions function at central and district-level decision-making activities.

FORMAL CENTRAL GOVERNMENT INSTITUTIONS

Ministry of Water Resources (MoWR) and Ministry of Physical Planning and Works (MPPW), through its district offices viz. Department of Irrigation (DOI) and Department of Water Supply and Sewerage (DWSS) respectively, are active government agencies at the district level. DOI is responsible for new development and/or rehabilitation of irrigation systems, while DWSS is responsible for the provision of drinking water and sanitation-related activities. These institutions activate themselves when the government provides budget for financial and technical services. MOWR, through the Department of Electricity Development (DOED), is directly involved in granting licences to the private sector for development of hydropower (e.g. the Indrawati hydropower project in this basin). The government has formed the Melamchi Water Supply Development Board (MWSDB) to transfer water from Melamchi Khola to Kathmandu Valley. The Water and Energy Commission Secretariat (WECS) has initiated focussed studies in the Indrawati river basin with an aim to analyse constraints and opportunities of managing water resources following the principles of IWRM.

DISTRICT-LEVEL INSTITUTIONS

There exists a district water resource committee (DWRC) in every district to register all the water use activities, as per Water Resources Act/ Regulations (1992/93), as well as to coordinate all water-related activities and help resolve the local water conflict within the district. The Chief District Officer (CDO) is the chairman of the committee and members almost entirely consist of water and other development-related district-level government officers. The DWRCs do not have representatives from local water users communities. Their function, so far, has been relegated to the registration of water users committees for irrigation systems.

The district development committee (DDC) is the government institution at the district level. It consists of elected representatives and is responsible for coordination and implementation of all local development activities. However, the DDC president of Sindhupalchowk district opined during an interaction meeting that 'the coordination between various agencies in the district for the development of water resources was lacking, since the central government in practice, does not recognize the role of DDC in the development of water resources in the district'. This is despite the fact that (a) that the Local Self Government Act provides authority to the DDC to utilize the water resources within the district to the benefit of the people of the district and (b) the DWRC is in existence (but not functional).

VILLAGE-LEVEL INSTITUTIONS

The village development committee (VDC) is the institution at the grass-root level, where all activities of the people related to governance system is initiated, deliberated, and acted upon. The Local Self-governance Act (1999) has provided authority to VDCs to function in eleven different sectoral categories whereas the Water Resources Act (1992) do not recognize them in specific terms. In practice, the VDCs are involved primarily in providing small financial support to VDC-level projects, including water sector activities like construction/improvement/ rehabilitation of irrigation systems, drinking water supply facility, watermills, community micro-hydro, as well as resolving conflicts over water uses.

NGOs

There are sixty-five NGOs officially registered in Sindhupalchowk district. The sudden increase in the number of registered NGOs is the outcome of Melamchi Inter-Basin Water Transfer Project, which has professed the principle of people's participation in the project. Though most of the NGOs are focussing their activities in the implementation of economic packages under the project's compensation programme, they have played an important role in raising awareness and concern among local people regarding the Melamchi Project and its likely impact on the local communities. NGOs have been trying to help the stakeholders to ask the right type of questions (to government authorities) regarding the project activities which affect their livelihoods, socio-economic, and cultural systems. The local communities are also advised by NGOs to seek remedies if it is perceived that it will adversely affect them.

WATER USERS ASSOCIATIONS

Formal water users associations (WUAs) necessarily get registered with the DWRC or the DDC primarily when they need external assistance. Many such WUAs exist in the Indrawati river basin, but as soon as the external funds start to dry up, their activities too seem to reduce to a considerable extent. Members of such associations again become active only when they have to mobilize external funds for major repair and maintenance. User committees (UC) are also required to be organized in other sectors like micro-hydro projects under the Rural Energy Development Programme (REDP). These UCs participate in all activities of the project and eventually take over the project for O&M. These conditions of setting up of UCs in different sub-sectors, with structured organizations wherein development and application of regulations are practised, are helping to enhance the local-level capacity to organize, plan, and manage development projects.

INFORMAL INSTITUTIONS

As elsewhere in Nepal, the Indrawati river basin has also age-old traditions and customs to manage the socio-economic system. The survival instinct of people has placed food security as the most important factor in traditional customs, values, and norms. In the water sector, therefore, the irrigation system owned by the community gets the first priority. Irrigation is important for food production, and activities of watermills and ghatta operation depend upon the success in producing agriculture crops in the region. Drinking water in most mountain and hill communities is made available from spring sources, hence drinking water in general does not interfere with other river water supply systems. The customary rule is formed by community leaders following traditional practices, common sense, as well as equity approaches.

Informal institutions are usually hidden deep in social customs and cultural traditions. Among these informal water use rules and institutions, the water allocation mechanisms between irrigation systems and the watermill/ghatta is particularly interesting and fascinating. For instance, the mill owners or micro-hydro committees will be responsible for rehabilitation, operation, and seasonal maintenance of irrigational canal, and water is available to farmers almost free of O&M cost. Similarly, the separation of any two intake sites by at least 200 m along the river means the conflicts are minimized as much as possible.

MELAMCHI WATER DIVERSION PROJECT

It is proposed that 1.97 m³/s (170 MLD) of water is diverted to Kathmandu valley from Melamchi Khola, a tributary of Indrawati River. This is an inter-basin water transfer project. The urban population of Kathmandu valley is experiencing drinking water shortage for a long time. The water diversion works involve construction of a 26.5-km tunnel in fragile Himalayan mountain environment. Donors have pledged to support the half billion-dollar project with conditions that the urban water supply system in Kathmandu Valley is privatized at operational level and water-related institutions are reformed so as to be able to properly address the broad water resources management issues of the valley before the project is launched.

Hydrological analysis in the basin, with limited data, indicates that the average dry season (March) flow of Melamchi Khola drops to a minimum of less than 3 m^3/s. There is a mandatory condition that at least 0.4 m^3/s is released to the downstream use in all seasons. These conditions and other secondary effects call for an exhaustive water balance study in the Melamchi sub-basin, so that proper and adequate contingency planning activities could be taken up.

Allocation of about four per cent of the total project cost (estimated at over US$ 460 million) for compensation package (not including compensation for land acquisition) is substantial. The compensation package is mostly targetted to provide public infrastructures rather than direct support to project affected people (PAP) like Ghatta owners and irrigation systems immediately downstream of the river diversion site.

Improved road access to Kathmandu as well as within Melamchi valley plus one fifteen-bed hospital and a higher secondary school are some of the attractive compensation packages offered to the Melamchi valley communities. Several local NGOs are involved in identifying critical issues and possible areas in which the Melamchi project could get involved so that proper compensation could be delivered to the Melamchi valley stakeholders.

East Rapti River Basin[3]

The East Rapti river basin is a part of Sapta Gandaki river basin (Figure 5.3). It lies south of Kathmandu valley and is contiguous to Bagmati river basin. The basin lies in the administrative districts of Makwanpur and Chitwan, covering 55 per cent of Makwanpur and 82 per cent of the total area of Chitwan district. The catchment area of the basin is about 3200 km^2. The East Rapti River originates in the high Mahabharat range of mountains (2400m above mean sea level) with steep slopes and it flattens out as it reaches the river valley and moves along the Chitwan valley to meet the Narayani River (Sapta Gandaki). The length of the main course of East Rapti is about 122 km. The main tributaries are Karra, Samari, Manahari, Lothar, Mardar, Pampa, Budhi, Kair, and Khageri Kholas.

WATER BALANCE SITUATION

Analysis of annual water balance in the basin indicated that net inflow of water in the basin for typical dry (1992), normal (1979), and wet (1978) years were 4565, 6120, and 7170 MCM respectively.

[3] Based on the study conducted by IAAS, Rampur, DOI, and IWMI/Nepal (2000).

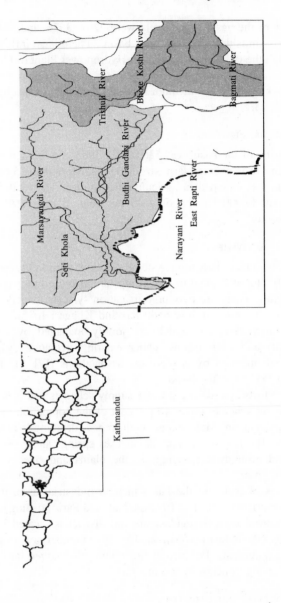

Figure 5.3: East Rapti sub-basin

Net outflow of the basin respectively were 2201, 3576, and 3848 MCM. Hence, the basin is described as 'open basin'. There is an inflow into the basin through the tail race of Kulekhani Hydropower Plant (60 MW) built in the adjacent Bagmati Basin. This water storage project has a catchment area of 126 km^2. The 114-m- high rock-file dam has a capacity to store about 83 MCM.

SOCIO-ECONOMIC FEATURES

The population of the basin was 536,000 in 1993 and the density is said to be 145/km^2. The average farm size is 0.9 ha. More than 75 per cent of the population is engaged in agricultural activities. The total cultivated area is about 86,000 ha and only 45 per cent of cultivated area has irrigation facility.

WATER USE ACTIVITIES

There are 214 surface irrigation systems. Shallow tube wells, (numbering around 580), dug wells, and treadle pumps are also used to irrigate in the Chitwan valley. There are two large irrigation systems namely Narayani Lift Irrigation (command area 8500 ha) and Khageri Irrigation System (command area 3900 ha) which are jointly managed by an agency (government) and user groups. Other existing irrigation systems are managed and operated by water users associations (WUA). There are around 200 WUAs in the basin.

There are forty-five drinking water supply schemes and it is said that the government schemes cover 40 per cent of the population in Chitwan and 43 per cent in Makwanpur with piped drinking water supply connections. The rest of the population uses springs and river water in the hills and groundwater sources in the plains and river valleys for domestic purposes.

Other uses of water in the basin include industrial uses, traditional watermills, tourism (boating in Chitwan National Park), fishing etc. In the case of industrial uses (Hetauda industrial district is located inside the basin), most of the industries (estimated at 139 in number) use groundwater for all its requirements. Polluted effluent being discharged into the river course is causing concern in recent times.

WATER SHORTAGE AND DISPUTES

Though the basin is termed in average as 'open basin', because of the spatial and temporal nature of rainfall, the basin experiences seasonal water scarcity typically during the dry months. Consequently, many

WUAs are making extra efforts to augment water supplies from alternative sources, including groundwater for irrigation as well as for drinking water use.

Conflicts between irrigation systems that draw water from the same river source are getting more frequent during dry season. Multiple uses of water are increasing and inter-sectoral water conflict is emerging. Some cases are listed as below:

(i) It is alleged that the limestone quarry site of Hetauda Cement Factory located in the upper reach of the Rapti river has aggravated the river sedimentation process, resulting in rising of the river bed which has effected changes in its low flow courses adversely affecting the intake capacity of irrigation systems downstream as well as the river regime.

(ii) Conflicts have been reported between the hoteliers in Royal Chitwan National Park (RCNP) and the irrigation system upstream of the park. Hoteliers complain that sufficient water for rafting and boating for tourists is not available during the dry season. Irrigators complain that the hotel workers dismantle their temporary diversion structures (brush weirs) at night adversely affecting the irrigation schedule.

(iii) Industrial effluents of some factories like textile, leather, soap, cement, and feed industries are discharged into the river without treatment. The downstream stakeholders and users like traditional fishermen, livestock farmers, as well as the wildlife and ecosystem of the national park are adversely affected. Conflicts are emerging.

(iv) District forest offices complain that the irrigation canal and drinking water supply alignment pass through hill forest area and sometimes cause or accelerate landslides and soil erosion, which leads to deforestation. These offices are demanding the proper design and management of water utility facilities, including the proper maintenance of canal alignment in the forest area.

WATER-RELATED INSTITUTIONS IN THE BASIN

As mentioned in the earlier case study, there are a number of government institutions involved in water management at a sub-sectoral level. User associations are formally established, not only to manage the water delivery system but also to establish their water rights. DWRCs, both in Makwanpur and Chitwan districts, have been registering WUAs in their respective districts but the issuance of licences to use water resources have not been operationalized.

WATER RIGHTS

Water rights in Nepal is in general acquired in the following four ways (Khadka, 1997):
   (i) Natural rights for developing water for a limited purpose;
   (ii) Rights acquired through licences for developing water resources for a specific purpose;
   (iii) Upper riparian has priority right compared to the lower riparian; and
   (iv) Customary use right and prior appropriation right.

Water rights are also associated with land rights. In irrigated lands, the rights to use are automatically transferred when the ownership changes. The prevailing practices of customary rights in the river basin are water-share based on investment, water rights purchased from others, and water rights proportionate to the land in irrigated areas.

WATER DISPUTE MANAGEMENT

Depending upon the nature and severity of conflict, a range of mechanisms is available to resolve disputes. Informal mechanism like negotiation between user groups is frequently practised to resolve the conflict. However, as time passes and as the water scarcity grows, conflict resolution will be sought through formal arrangements. If the dispute falls in the territorial jurisdiction of a VDC, the chairman of the VDC seeks negotiation between disputing parties and if it fails then the chairman acts as an arbitrator and gives his verdict, which in most cases are adhered to.

There are no institutional arrangements to cope up with the multiple uses of water. DWRC is expected to address such issues in future. DWRC needs to be made functional with offices of their own, including staff with adequate experiences in water management.

KANKAI-MAI RIVER BASIN

Nepal Water Partnership (NWP), an NGO, has identified Kankai Mai river basin to develop it as 'an area water partnership' (AWP) for proper water resources management. Its aim is to provide a platform to all water-related institutions and stakeholders in the basin for interaction to deliberate and solve the emerging issues related to water use so that IWRM can be achieved at local level (Figure 5.4).

The formation of AWP is one of the modalities put forward by the South Asia Technical Advisory Committee (SASTAC) of Global Water Partnership (GWP), to operationalize the concept of IWRM in water resources management.

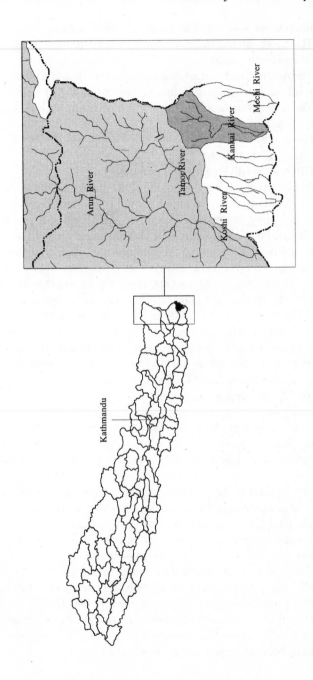

Figure 5.4: Kankai-Mai river basin

The reasons for selecting this basin for AWP are:
  (i) The entire catchment lies within the administrative boundary of an administrative district.
  (ii) The water in the basin is being used extensively and it is expected to result in water stress situation in the not-too-distant future.
  (iii) The area has an adequate level of infrastructure such as roads and communication facilities.
  (iv) Government and NGOs are working closely with community-based organizations (CBOs).
  (v) CBOs are relatively proactive in nature.

Kankai Mai River originates from the Mahabharat range at an elevation of about 3300 metres above mean sea level. For the purpose of the study, the lower boundary of the AWP basin has been fixed at a place called Chepte (300m above mean sea level). The catchment area of the basin is 1150 km^2 and lies wholly within the administrative district of Ilam.

Kankai Mai receives flow contributions from the following main tributaries: Mai Khola, Jog Mai Khola, Puwa Khola, and Deb Mai Khola. The gauging station close to Chepte has shown the maximum instantaneous flow of 7500 m^3 (12 August 1990) and minimum instantaneous flow of 3.20 m^3 (29 May 1987). The basin can be termed as an 'open basin'.

The catchment area covers 68 per cent of the district and includes all major urban and semi-urban areas. The estimated population in the catchment area is about 200,000. The population is expected to grow at the rate of 2.56 per cent per annum.

Tea production followed by ginger and cardamom are the major cash crops of the basin area. Most of the hill slope area is covered by tea shrubs and forest in general. Paddy is grown in only about 5 per cent of the catchment area, less than half of which has irrigation facility at present.

With the increase in the population in the river basin, the consumptive use of water is expected to increase. The cultivable area of around 60 per cent of the total land area will demand more irrigation water in future. Potential water use conflicts during dry periods have started to emerge.

The key stakeholders are: (a) government line agencies and public corporations; (b) local government bodies including DWRC; (c) private/ community micro-hydro plant owner/operator (there are 190 micro-hydropower plants in the catchment area, three of which, with a capacity of more than 5kW, are considered large); (d) non-government organization; (nine active NGOs have been identified); and (e) WUAs (five active WUAs have been identified as key stakeholders).

Based on the preliminary baseline report, a stakeholder workshop was held in Ilam on 20 May 2002. About forty-five participants from different institutions including representatives from government, local elected bodies, and NGOs took part in the workshop. The participants decided to form the Mai area water partnership with Namsaling Community Development Centre (NCDC), a local NGO, as the local coordinating institution. A committee to draft the 'constitution' of the partnership was formed and the constitution so drafted was agreed upon by the stakeholder in a meeting held on 8 August 2002. The application to register the 'partnership' has now been filed at the office of chief district officer (CDO) of the district of Ilam. Once the 'partnership' is legally established, it is expected to operate and influence the water-related activities in the basin so that the principles of IWRM could be applied in the water resources management of the Mai basin.

Upper Bagmati River Basin

His Majesty's Government of Nepal has initiated, with the assistance of Asian Development Bank (ADB), the project called 'Optimizing Water Use in Kathmandu Valley' which is in effect an exercise in integrated water resource management for the sustainable development of water resources in Kathmandu valley, which comprise the Upper Bagmati river basin. The Kathmandu valley had been facing an acute shortage of drinking water supply in the municipal areas for quite some time. The population of the districts in Kathmandu valley adds up to 1.66 million according to the recent census (2001). The project commenced on 11 March 2002 and was completed as per schedule in March 2003.

The original objective was focussed on the development of computer models and train counterpart staff to help in developing and managing water resources within Kathmandu valley on a long-term sustainable basis. This was expanded later to include preparing an action plan for the near term.

One of the major activity of the project was to develop a 'database for water management in Kathmandu valley'. The collection, collation, validation of data related to water was not an easy task, in the present inward looking bureaucratic set up.

Data were categorized into different groups. Meteorological data contained data on rainfall, temperature, and evapo-transpiration. River flow data contained data on river discharges and river quality. Groundwater data contained data on wells, water quality of wells and spouts. Water use data contained data on irrigation and domestic uses. Supply centre and source data included information supply centres and sources of supply.

Demand centre data included zoning of demand centres. The primary purposes of the models among others are to:

(i)   Determine the effects on the system including existing users, of any new projects, including new intakes, wells, surface storage, recharge areas either by wells or infiltration areas.

(ii) Determine the conservation capacity of catchment to increase water recovery including roof-top collection and drainage, and collection of run-off from low permeable surface for use.

(iii) Assess the increase usability of water arriving from wastewater treatment as well as in-stream requirement.

(iv) Evaluate the change to water supply situation arising from watershed management and forest protection.

(v)   Identify through stakeholder consultation process, the most appropriate mix of future projects to improve the present situation.

(vi) Examine the anticipated impacts on demands that would arise from changed pricing strategies including water rights trading.

The simulation models (MIKE SHE and MIKE BASIN) and an optimization model are proposed to meet the purposes.

FORMAT OF STAKEHOLDER CONSULTATION

Another important feature of the project is the design of a series of different stakeholder consultation processes. Attempts have been made to make stakeholder consultation an integral part of the process of developing a consensus-based water resources management system.

A stakeholders' consultative committee, consisting of about twenty-five persons from different government and non-government organizations, professional societies, water user associations, and local government bodies, including politicians was constituted. The meeting of the consultative committee was planned to be held frequently, if possible in a monthly basis (four such meetings have been conducted). The responses from committee members were quite encouraging.

Three stakeholder workshops of 80 to 100 persons, representing broad stakeholder constituency, were planned (two of them have been adequately completed).

Discussion sessions with focussed target groups were executed. The focussed target groups included NGO Forum for Kathmandu Valley Water Supply, women groups; farmers' groups, carpet associations and local governments.

Appeals to the public, soliciting comments and information, were made in newspapers as well as over the electronic media. However, the responses were feeble.

It will be interesting to monitor the outcome of the study as well as the processes involved in the implementation of the recommended options of the project. The developed computer models as tools for integrated water resources management for Kathmandu valley is the first of its kind in the country and will have far-reaching implications if the application of this decision support system is successfully achieved.

Conclusions

It has now been recognized that though Nepal has an abundance of water resources, the country faces water shortages during dry periods of eight months in a year because of the spatial and temporal nature of the precipitation. Certain levels of water stress during dry periods is imminent, unless the available water is well managed.

The framework of IWRM planning has received wide acceptance for effective water resources management. Nepal has some legislative framework in place to initiate integrated water resources management but a lot more still needs to be done including harmonization of different legislations for these to be more effective. The activities to improve the institutional and legal environment are going on through the preparation of the national water plan.

Water conflicts are resolved, at present, using customary and traditional rules and practices. These customary rules alone will not be sufficient to resolve disputes arising out of water scarcity in future. Participation of all stakeholders during negotiations pertaining to allocation and management of water resources will emerge as crucial in the future. Special efforts must be made to include all stakeholders (specially the poor and marginalized population) in the process of equitable distribution of resources to avoid/ minimize conflicts.

Institutional arrangement is a necessary to operationalize the IWRM principle. In this context, the District Water Resource Committee needs to be made more functional and active at the local level and representation in DWRC should be broad based by including user committee members from among the stakeholders. DWRC should be properly manned, trained, and equipped to manage the water resources within the district, keeping in view the principles of IWRM principles. DWRC must have its own secretariat with experienced staff in water management to be functional and effective.

Environmental awareness relating to river water pollution is increasing and so is the inter-sectoral conflict. There is a need to enforce the existing regulatory mechanisms by involving local government institutions and stakeholders. DWRC should also be entrusted with the regulatory function, and this should be clearly stated in its mandate.

The concept of water rights and water trading needs to be debated extensively among all stakeholders. Clear and detail mechanisms must be established so that these issues of water rights and water trading do not hinder the development of water resources.

At the central level, an institution like Water and Energy Commission Secretariat should be mandated to oversee all developments related to water resources so that the related development follows the principles of integrated management. It must be recognized that the process of development is a dynamic one and the central body should have a mechanism to review all regulatory and legal instruments in a fixed interval of time. A formal link must be established between the Water and Energy Commission Secretariat and the DWRC for IWRM to function properly.

References

Adhikari, K.R., M. Bhattarai, Major Issues and Concerns in East Rapti and Indrawati River Basins: A Comparative Analysis', Paper presented at a Policy Dialogue Meeting on Basin Level IWRM in Nepal, Kathmandu: IWMI, 2002.

Biswas, A.K., and J.I. Uitto (eds), *Sustainable Development of the Ganges-Brahmaputra-Meghna Basins*, Tokyo: The United Nations University, 2001.

Global Water Partnership; Integrated Water Resources Management; March 2000, Denmark.

'Integrated Development and Management of Water Resources, A Case of Indrawati River Basin Nepal', Proceedings of Workshop, Kathmandu, 25 April 2001, WECS and IWMI, Sept. 2001.

Karki, A., 'Kankai-Mai Khola River Basin Study', mimeo, Kathmandu: Jalsrot Vikash Sanstha/Nepal Water Partnership, 2001.

Kayastha, R.N., D. Pant, 'Institutional Analysis for Water Resources Management in East Rapti River Basin, Nepal', mimeo, International Water Management Institute, 2001.

'Optimizing Water Use in Kathmandu Valley Project', Inception Report of Acres International i.a.w. Arcadis, East Consult, and Water Asia, Ministry of Physical Planning and Works, His Majesty's Government of Nepal (ADB TA 3700–NEP), 2000.

Shrestha, T.N., *The Implementation of Decentralization Scheme in Nepal: An Assessment and Lessons for Future*, Kathmandu: Joshi Publications, 1999.

Water Resources Strategy, Nepal, Kathmandu: Water and Energy Commission Secretariat, January 2002.

6

INSTITUTIONAL SET-UP FOR INTEGRATED MANAGEMENT OF THE KLANG RIVER BASIN, MALAYSIA

Keizrul Abdullah

Introduction

Malaysia is a country rich in water resources. Located in the humid tropics (see map in Figure 6.1), the climate is equatorial and is influenced by the north-east and south-west monsoons. The former, prevailing between November and February, brings heavy rainfall (as much as 600 mm in twenty-four hours) predominantly to the east coast of Peninsular Malaysia and to Sabah and Sarawak. Rain-bearing winds also come with the south-west monsoon from April to September though rainfall during these periods are generally less than during the north-east monsoon. Together, these two monsoons bring an average annual rainfall of 3000 mm.

The water resources in Malaysia (Government of Malaysia [GOM], 1982) are summarized in Table 6.1.

The water demand for the past three decades (GOM, 2002) are given in Table 6.2 (the values within the brackets refer to the proportions of the total water use). Table 6.3 shows the national water supply production capacity and coverage and non-revenue water (NRW) for 1990 and 2000.

Table 6.1: Water resources in Malaysia

Annual rainfall:	990 billion m^3
Surface runoff:	566 billion m^3
Evapo-transpiration:	360 billion m^3
Groundwater recharge:	64 billion m^3
Surface artificial storages (dams):	25 billion m^3
Groundwater storage (aquifers):	5000 billion m^3

Figure 6.1: Malaysia (Klang river basin shown in circle)

Table 6.2: Water demand for 1980, 1990, and 2000

(in billion m³)

Water user sector	1980		1990		2000	
Domestic and industry	1.3	(18%)	2.6	(20%)	4.8	(23%)
Irrigation	7.4	(80%)	9.0	(78%)	10.4	(75%)
Others	0.2	(2%)	0.2	(2%)	0.3	(2%)

Table 6.3: Water supply production capacity and coverage, and non-revenue water for 1990 and 2000

Item	1990	2000
Production capacity	6103 mld	11,800 mld
	(2.2 billion m³)	(4.3 billion m³)
National coverage	80%	95%
Urban coverage	96%	99%
Rural coverage	67%	83%
Non-revenue water	43%	38%

Due to the rapid population increase and the rapid growth of industries, the annual water demand for the domestic and industrial sector has been expanding at about 12 per cent.

Irrigation accounts for three-quarters of the water utilized. Irrigation development caters primarily for the double cropping of paddy to meet the dual objectives of increasing food production as well as raising the income levels of the farmers. Irrigation efficiencies, however, are low, reaching about 50 to 60 per cent for the larger schemes while some of the smaller schemes are operating at efficiencies of less than 40 per cent.

While the national coverage for drinking water has reached 95 per cent, sanitation coverage lags behind and in 2000, only 79 per cent of the urban population have access to central sewerage system though 98 per cent of the rural population have been provided with pour-flush latrines.

Rivers are the main source of water supply, contributing some 97 per cent of total supply with the remaining 3 per cent coming from groundwater sources. Pollution loads have increased over the past few years, adversely affecting the riverine environment and rendering river water unfit for use. The main sources of organic water pollution are domestic and industrial sewage, effluents from agro-industries, and animal husbandry. In several urban and industrial areas, organic pollution of water from both point and non-point sources have resulted in environmental problems and adversely affected aquatic lives.

Data compiled by the Department of Environment (DOE), points to a trend of a slow but steady deterioration in the water quality of the sampled rivers (see Table 6.4).

Due to the abundance of rainfall, water was not perceived to be a major factor in the socio-economic development of the country during the past decades. Lately, however, as a consequence of the rapid pace of development, the water situation in a number of river basins has changed from one of relative abundance to one of scarcity. Population growth and the expansion in urbanization, industrialization, and irrigated agriculture are imposing rapidly growing demands and pressure on the water resources, besides contributing to the rising water pollution. Water management in these basins is becoming increasingly comprehensive and complicated as society develops and the expectations of the people increase. One such river basin is the Klang River Basin.

Klang River Basin

The Klang River originates in the highland areas about 25 km north-east of Kuala Lumpur, the capital of Malaysia. The river, with a length of about 120 km (its eleven main tributaries add to another 700 km) flows through the Federal Territory and parts of the State of Selangor, before discharging into the Straits of Malacca (See Figure 6.2).

This river basin is the most developed and most densely populated area of the country. With a catchment area of slightly less than 1300 km^2 (roughly twice the size of Singapore), it supports a population of about 3.7 million people (18 per cent of the nation's population). Development has averaged 5 per cent annually over the last ten years and the present land use is dominated by urban residential areas (44 per cent), followed by forest reserves (34 per cent), agriculture (15 per cent), and commercial and industry area (7 per cent).

BACKGROUND

Being the most economically active region of the country, the Klang river basin is facing an array of problems associated with managing water resources within the basin. The main problems faced are flooding, serious degradation in water quality, a depleted river corridor environment, development within the river channel, encroachment of river reserves by illegal settlers and factories, solid waste pollution, suspended sediment solid waste washed by land development works, and the occasional occurrence of water scarcity. These problems can be attributed to factors associated with the traditional way of managing resources within a river

Table 6.4: Quality of river water, 1993–99

Category	1993		1994		1995		1996		1997		1998		1999	
	No.	%	No.	%	No.	%	No.	%	No.	%	No.	%	No.	%
Very polluted	11	9.50	14	12.10	14	12.20	13	11.20	25	21.40	16	13.30	13	10.80
Slightly polluted	73	62.90	64	55.20	53	46.10	61	52.60	68	58.10	71	59.20	72	60.00
Clean	32	27.60	38	32.70	48	41.70	42	36.20	24	20.50	33	27.50	35	29.20
Total	116	100	116	100	115	100	116	100	117	100	120	100	120	100

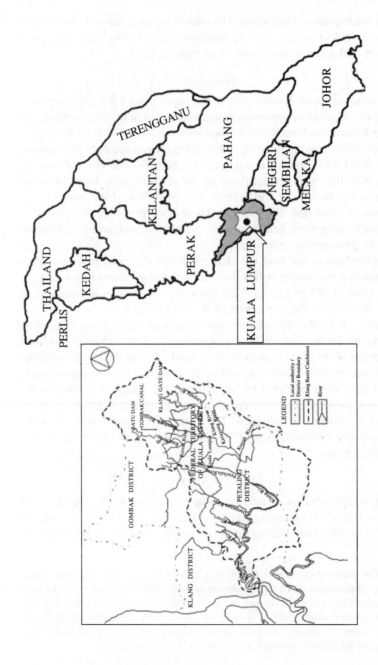

Figure 6.2: Catchment boundary of Klang river basin

basin, inadequate institutional and legislative set-up, a failure to encompass resource and environmental management in a holistic and integrated manner, and a lack of enforcement of existing laws.

FLOODING

As a result of rapid development in the upper part of the catchment and a lack of oversight in approving development within the river section, the Klang River is no longer able to cope with the increased flow discharges during storms, making flooding of low-lying areas a common phenomena. Since April 2000, the city centre itself has been flooded not less than six times with the mean annual flood discharges increasing from 144 m^3/s between the periods of 1910–86 to 440 m^3/s from 1987 onwards, an increase of more than 300 per cent. These flood events caused extensive damage to properties, hardship to the population, and severe embarrassment to government agencies.

An analysis of the land use changes within the basin reveals a large increase in urbanization, with a five-fold increase in urban areas within ten years from 12,461 ha to 67,592 ha. Most of these changes in urbanization were observed in the upper part of the catchment.

Since 1971, the government has embarked on a flood mitigation programme for the basin. A large proportion of the river channel was lined and enlarged to provide a protection level of 1 in 100 years average recurrence interval. However, within the city centre, river improvement works have been restricted by limited reserves. Further widening of the river will be very expensive due to the high degree of urbanization. To further aggravate the situation, the Klang river meets with its main tributary, the Gombak river at this point, making it vulnerable to any flooding event.

WATER QUALITY

Water quality remains a difficult issue to address for the Klang river system. Records from twenty-five water quality monitoring stations shows that the overall water quality index[1] (WQI) has been fluctuating between

---

[1] The water quality index (WQI) is a classification system incorporating a number of parameters such as ammoniacal nitrogen (NH3-N), biochemical oxygen demand (BOD), chemical oxygen demand (COD), dissolved oxygen (DO), pH, and suspended solids (SS). It is used to give a quick indication of the water quality using a scale of 0 to 100, where the index range 81–100 denotes clean water, 60–80 slightly polluted water, and 0–59 polluted water.

47 and 56 over the last ten years, indicating that the main stretches of the river remains polluted, with only the upper part of the catchment still fairly pristine. Part of the reason stems from the rapid development within the basin, where the tremendous increase in impervious areas has led to large flood run-off discharges and decreased base flows. Thus, in times of low flow, there is insufficient quantity to help in diluting the huge loads of wastes coming into the river systems.

The main sources of river water pollution are organic load and ammoniacal nitrogen from domestic sewerage. Many sewerage treatment plants are old, poorly maintained, and inadequate to cater for present loads, and as a result, treatment plant effluents exceed the allowable discharge standards.

In addition, pollution from non-point sources (NPS) are believed to significantly affect the quality of river water and contribute to river pollution though there has not been any study to quantify the exact pollution load produced by these sources. In Kuala Lumpur city alone, there are close to 12,000 food stalls which are not connected to the sewer system but discharges directly into the drainage systems. It is believed that the amount of NPS pollution is approximately equal to that of partially treated wastewater from domestic source.

As is common with cities in the region, Kuala Lumpur has its share of squatters' colonies that have mushroomed along river reserves. These colonies aggravate pollution by contributing to wastewater, sullage, and solid waste discharges into the river system.

SUSPENDED SEDIMENTS AND EROSION

A major problem in the Klang river basin is soil erosion and sedimentation. Soil erosion rate is high by both Malaysian and international standards and is estimated at 18 tons/ha per year or approximately 2.3 million tons of soil loss in the basin. From this, some 7.8 tones/ha or approximately 1.0 million tones of soil is washed into the river system each year. The major contribution to this is from indiscriminate land development works.

Another source that contributes to the high sediment levels in the rivers is stream bank erosion, and the total loss from the unlined section of the rivers is estimated at about 200,000 tons annually. Such excessive erosion and sedimentation levels have resulted in damages to drainage structures, sedimentation of river channels and the river mouth, reduction in the flood conveyance capacity of the river, and deterioration in water quality.

SOLID WASTES

Illegal disposal of solid wastes into waterways is another issue besieging the Klang river system. A recent study indicated that between 170,000 to 280,000 tons is estimated to enter the river system annually. Of these, only about 25,000 tons were estimated to be retrieved. The main sources of river solid waste originate from villages and squatters' colonies (40 per cent), illegal dumping (15 per cent) and from non-serviced areas (10 per cent).

The current methods employed to trap and remove solid wastes from the river system are far from efficient and as such quality of river water continues to degrade. The estimated cost to remove rubbish from the Klang River and its tributaries is US$ 2.1 million/year.

WATER DEMAND

Concurrent with its development, the population it supports, and with over 1500 major industrial premises, the Klang river basin has a very high demand for water supply. Over the last decade, there has not been any acute shortage in water supply except for the year 1997 when prolonged water scarcity affected the whole basin. However, the projected increase in population (by growth and migration) will push the water demand to a level close to the present capacity of the treatment plants, and new supply sources will be needed over the coming decade.

The recent national water resources study reported that surface water resources for Selangor state are close to full development with regulation of five existing storage dams in the Klang valley. The remaining potential for further development is in the neighbouring Selangor River catchment area. The Selangor dam project has started and this dam would be able to meet the demand until the year 2010. Beyond that, the basin and Selangor state would have to source its water supply through water transfer from Pahang state.

LEGAL AND INSTITUTIONAL ISSUES

LEGISLATION

There is a broad range of legislation that relates, or could relate, to river basin management generally and, as Malaysia is a federation of thirteen states, it is necessary to first understand the constitutional arrangements and the respective jurisdictions of the federal government and the states.

The Malaysian Constitution specifies the legislative power of the federal and state governments in the Ninth Schedule where three lists are provided. Parliament may enact laws with respect to any of the matters specified in the federal list (List I) or the concurrent list (List III). States may enact laws with respect to any of the matters specified in the state list (List II) or the concurrent list.

Water is more extensively specified in the state list. A direct reference to water occurs in item 6 of the state list that allows a State to legislate on water (including water supplies, rivers, and canals), control of silt and riparian rights. The state list also goes on to note other matters directly connected with basin management including agriculture, forestry, local government, land, land improvement, and soil conservation. Rivers entirely within a state are excluded from federal jurisdiction.

Drainage and irrigation is specified in the concurrent list where both the federal and the state governments have legislative power. Only when interstate issues or transfers arise or when no agreement exists between the states concerned, will rivers and canals come under federal jurisdiction. The state can also regulate catchment areas in relation to protection of water quality in rivers.

Thus, in general, water and basin management are state matters. This includes rivers, lakes, streams, and water beneath the surface of the land. However, the federal government has some specific powers over water supplies, rivers, and canals, where there are federal works present. In addition, the federal parliament may also make laws with respect to any matter in the state list for promoting uniformity of the laws of two or more states.

INSTITUTIONAL ISSUES

The Klang river basin is divided under a number of administrative bodies, viz. the Federal Territory of Kuala Lumpur, and parts of the Hulu Langat, Kuala Langat, Gombak, Sepang, Petaling, and Klang districts in the Selangor state. In addition, there are the municipalities for the towns of Klang, Shah Alam, Subang Jaya, Petaling Jaya, Kajang, Ampang Jaya, and Selayang.

Given the diverse and, at times, overlapping roles of the various agencies involved, river management for the basin is complex. The main players are the Selangor water management board, Kuala Lumpur City Hall, Department of Irrigation and Drainage (for the Federal Territory and Selangor), Department of Environment, and the respective local authorities (local governments). To further complicate matters, there is a Klang valley Planning Council, a consultative body comprising of both federal and state-level agencies, and at the national level, a National Water Resources

Council (NWRC) was established in 1998 to provide high-level direction and policy, following the water crisis experienced in the country due to a serious drought attributed to the El Niño effect.

The Klang Valley Planning Council was established in 1981 to address the development of the Klang valley during the period of significant growth during the 1980s and 1990s. The council is a high-level committee chaired by the Prime Minister, and looks at regional strategic and development plans and the review and evaluation of other matters that may affect development of the region. Its mandate covers a wide range of issues, including transport, land use, and infrastructure, but it does not include flood mitigation and its links with land use and/or long-term planning. In addition, the council is unwieldy, having too many members, both political and technical, and in its current form has not been too effective.

The council is supported by a working committee which includes all the local governments in the Klang valley, and with relevant federal and state departments attending as required. Originally, the working committee was co-chaired by the Selangor state secretary (the senior-most civil servant in the state of Selangor) and the mayor of the Federal Territory of Kuala Lumpur, but this arrangement has very recently been changed with the appointment of the Chief Secretary to the federal government as chairman. The responsibilities of the Committee include certification of policies and strategies for approval by the council, ensuring their effective implementation and coordinating, monitoring, and evaluation of the subsequent implementation.

The Committee does not have an adequate executive role when compared to the model of, for example, the Murray-Darling Basin Commission. It is not geared to initiate investigations or projects, nor can it coordinate the activities of the various departments with their specialized, even compartmentalized roles and responsibilities. A key weakness in the institutional set-up is the rolling membership of government agencies, depending on the issue for discussion.

Both the Council and the Committee are supported by the Federal Territory Development and Klang Valley Planning Division (FTDKVPD), located within the Prime Minister's department. The main responsibilities of the FTDKVPD include monitoring development in the region, amending policies and strategies to be certified by the committee and approved by the council, evaluating plans, and maintaining a database of the region. It is also required to assist in overcoming issues in implementation of development projects as well as having a coordinating role in transport, environmental issues, and in resolving issues between Selangor state and the Federal territory.

The FTDKVPD is not adequately resourced to undertake effectively these broad-ranging tasks. It has not been able to initiate activities beyond its limited framework and, if it is to be effective in its role, it must be strengthened in both resources and powers. It is also very heavily 'planning' biased, with only a limited view of the meaning of 'integrated river basin management'. All too often, IRBM is seen as river management only, with no relationship to land use and/or planning. Without an understanding for and appreciation of the long-term benefits of integrated planning for all resources in a river basin, the existing problems are likely to get worse and the future problems will simply add to the existing.

Integrated Water Resources Management: The Emerging Water Scenario in Malaysia over the Ages

In recent years, there have been persistent calls on the need to manage water resources in an integrated manner through a process called integrated water resources management (IWRM). The concept of IWRM itself is not new, having been discussed and recommended at the Mar del Plata Conference in 1977, and it formed one of the four principles from the Dublin Conference in 1992. Whilst there are many ways of defining IWRM, one concise definition is the one adopted by the Global Water Partnership (GWP) which has defined IWRM to include:

(i) The evaluation of the quantity and quality of available water resources under alternative land uses;

(ii) The allocation of raw water and reused water to competing uses and users; and

(iii) The development of water supply and demand management strategies and mechanisms to increase welfare derived from scarce resources of water and capital in a sustainable manner.

In Malaysia, this concept is rapidly gaining prominence as a result of emerging problems and issues relating to water resources, particularly that of water shortages, flooding, and deterioration in the quality of river water. This is in stark contrast to the situation in the early 1900s when water was regarded as an abundant resource.

THE EARLY 1900S

At the start of the last century, Malaysia (then Malaya) was basically an agrarian society with rubber and tin as the country's main exports. As water was easily available, there was no concept of managing water in an integrated or holistic manner. Water was managed individually by the various sectors, the two main sectors being domestic water supply and irrigation. Water schemes were available only in some of the bigger towns, while irrigated agriculture depended mainly on rainfall. The food crisis in 1918–20 and escalating import prices prompted the government to review the rice deficient situation and, in 1932, a new department was formed to provide irrigation and drainage facilities for increasing food production. With the abundant rainfall, the major problem was drainage rather than irrigation, and the new department was given the name of Drainage and Irrigation Department. The main dimensions of water management was on technical and financial matters.

THE 1960s

When Malaysia gained independence in 1957, there was a shift of emphasis to the need to improve the socio-economic status of the rural poor. A rural water supply programme was initiated to provide potable water to the villages, while urban water supply schemes were intensified. As the rural population was mainly agricultural, the government embarked on irrigation projects to not only increase food production, but also as a vehicle for rural upliftment. One of the main methods was by increasing the production of rice from one to two crops in a year, with the additional second crop providing the opportunity for increased farm income.

The earlier irrigation systems, however, could not cope with the demands of double cropping as larger amounts of water were required for the dry-season crop. The irrigation system, therefore, had to be upgraded to cater for both supplementary supply during the wet season and full supply during the dry season. Water resources development became an important component of irrigation projects, and storage dams, barrages, and major pumping stations were constructed. Irrigation canals were upgraded to meet the increased water demands for the off-season crops.

Water resources management, however, remained very sectoral though conflicts in water use were beginning to arise.

THE 1970s

Throughout this period, the twin objectives of poverty alleviation and food production continued as before. However, with the general completion of the double cropping programme, yield increase per unit area per season was the next option to meet the above objectives and thus maintaining irrigation sustainability. Thus, irrigation was slowly shifting its emphasis from the structural aspect to *water management issues*. In order to manage the water resource, rotational supply became necessary and therefore scheduling was introduced. In addition, to maximize yields and conserve water, new water management practices were introduced to meet the crop water requirement for the different stages of growth. This required the intensification of irrigation infrastructure, and the introduction of regulating and measuring devices.

Around this period, the country embarked on an industrialization programme which resulted in a gradual shift in population from rural to urban areas. As population concentrations grew, so too did competition for water. Towards the end of this period (in 1978) the national water resources study (GOM, 1982) was initiated with the objective of determining the country's water resources and to prepare a strategic plan for the future. Management of water was still on a sectoral basis, but communication between sectors was established through administrative mechanisms such as inter-agency committees. Master plans prepared during this period were mainly single-sector based.

THE 1980s

The 1980s was a period of major challenges for water resources management. The country's policy on industrialization was beginning to show rapid progress and the manufacturing sector became the engine of growth for the economy. Water supply to industry and to the accompanying residential areas grew at an exponential rate. Pollution of rivers and waterways became a major problem and during the drier seasons, water shortages caused disruptions to the supply.

The most significant impact on irrigation was the shift of labour from agriculture to industry and consequently more and more lands were left idle. The earlier concept of upliftment through irrigation development became more untenable as industrialization took root. A new national agricultural policy was developed which specified that food production was to be concentrated in eight large granary areas. The farmers in these areas gradually adjusted to the situation through the adoption of labour-saving practices such as mechanization and replacement of transplanting by direct seeding. While it helped to overcome labour shortages, it created new problems as it required good on-farm water management and levelled fields. Peak irrigation water demand increased, causing more stress on the available water resources.

During the period, there were a number of droughts. Administrative rules were drawn up to determine the priority on water use. Discussions on the need for policies and legislation on water resources management began and efforts were made to include the needs of other sectors in the preparation of master plans and studies.

THE 1990s AND EARLY 2000s

Towards the end of the last century, the country experienced a number of El-Nino events which resulted in more extremes in rainfall patterns. This led to a series of water shortages, flooding, and landslides. Following on from Dublin and Rio (UNCED, 1992), and the growth of civil society, there was greater awareness among the public on water and environmental issues, and through the efforts of environmental NGOs there were increased calls for the introduction of IWRM.

A nationwide 'love our rivers' campaign was launched in 1992 and IWRM was introduced into the seventh Malaysian Plan (a five-year development plan covering the period 1996–2000) document. Privatization of water supply projects became common and a national privatized sewerage programme was started. After a particularly serious water crisis in 1998, a National Water Resources Council was formed, with the Prime Minister as its Chairman, and a membership which includes the Chief Minister of all states, the Finance Minister, and the Agriculture Minister, among representatives of other ministries.

Around this time, there were numerous forums and discussions on IWRM. The concept of IWRM became widely accepted among the water professionals, but there were problems in making it a reality. Under the Malaysian Constitution, land and water falls under the jurisdiction of the states. This has made efforts to manage water resources in an integrated and holistic manner difficult. For example, the current efforts to develop a national water policy have yet to receive the concurrence of all the states. To make the IWRM concept more easily understandable, the linkage of land and water interactions within a river basin was promoted, leading to the concept of integrated river basin management (IRBM). This was then reflected into national policies through the Third Malaysian Outline Perspective Plan (OPP3, 2000–10) and the Eighth Malaysian Plan (2001–5).

IRBM

A river basin is the area delineated by its natural hydrological boundaries such that any rain falling on the area will drain to the river. Thus, any activity that takes place upstream in a river basin will eventually have an impact on the quality and quantity of the river water as its reaches the downstream. The opening up of land, logging, industry, and mining are examples of activities that will significantly affect rivers. Industries incorrectly located will affect the water quality in the river. Developments and townships need to be properly planned to preserve the natural beauty and functions of the river. Integrated river basin management is the holistic approach to managing the river basin with the objective of protecting and preserving the river and its ecosystem.

IRBM can be defined as 'the coordinated management of the resources existing in the natural environment, comprising of air, water, land, flora, and fauna, based on the river basin as a geographical unit, with the objective of balancing the needs of man to utilize the resources for the improvement of his living conditions with the necessity of conserving the resources to ensure their sustainability'.

IRBM is geared towards integrating and effectively coordinating policies, programmes and practices addressing the water and river-related issues while trying to balance the needs of socio-economic development with the needs of conservation and protection of the environment. The water-related issues include the efficiency of water use, long-term resource protection, and the economic effects of deterioration in water quality; while the river-related issues include river basin-based management of water resources and wastewater, data collection and dissemination, and model development. This process will require improved professional capability, and increased financial, legislative, managerial, and political capacity.

An institutional framework is necessary to put into IRBM into practice. The word 'institutional' has to be seen in a wider context to include not just government, but all stakeholders, including the private sector, NGOs, and the general public. The framework will lay out the roles and functions of the various institutions at different level.

IRBM APPROACH AND INSTITUTIONAL REFORMS

In formulating solutions to see how resources in Klang River basin can be managed in an integrated manner, it is important to determine an effective, long-term institutional arrangement for integrated river basin management in the Klang Basin. This objective is not confined to single-focus issues such as pollution and flood control but involves the development of a sustainable interaction of human activities with natural resources so that development and protection of the environment are harmonized.

These activities include:
  (i) Land use planning;
  (ii) Water quality and quantity catchment protection;

(iii) Water harvesting and wastewater disposal;
(iv) Sustainable forestry practices;
(v) Flora and fauna conservation;
(vi) Reserving sensitive areas including river banks, floodway and detention areas, wetlands, highlands, recharge and groundwater areas;
(vii) Flood storage areas;
(viii) Coastal swamps; and
(ix) Environmentally sensitive areas.

Integration and actual implementation of all these activities is vital to achieving sustainability.

CONTEMPORARY PRACTICES

There is no 'right' model that will suit all countries or states within a country or a local region or area to achieve efficient and effective integrated river basin management. Thus, in developing an institutional model for the Klang river basin, the overarching concept was to examine 'best practices' from around the world, and to incorporate those which are suitable and acceptable by the stakeholders in the basin. The eight guiding principles for sustainable development that was established at the UNCED Earth Summit (1992) were felt to be most suitable and relevant and have been used as the basis for the Klang river basin. These principles are:

(i) River basin-based strategies: The best practice in managing rivers and water resources is based on geographical river basins or water catchments. This concept has been successfully practised in the United Kingdom, most of the other European countries, United States of America, Australia, and is now emerging in south-east Asian countries.

(ii) Towards sustainable development: Overcoming the problems of depleting water resources, droughts, floods sedimentation and pollution, through water conservation and responsible water management. The principle of sustainable development ensures the protection of the natural resource in terms of quantity and quality for present and future needs and includes water demand management and ecosystem conservation. It embodies the need to establish the relationships between human activities and land use with environmental protection.

(iii) Integrated and multifunctional approach: An integrated and multifunctional approach ensures the availability of resources for all users through coordination and cooperation.

(iv) Separation of the functions of regulatory mechanisms and service providers: The separation of these functions avoids conflicting interests and allows transparency of actions. Water suppliers, effluent dischargers, and flood controllers cannot be policymakers and regulators at the same time. The establishment of business enterprises and assigning roles to different areas of government or privatization programmes helps to support this principle.

(v) Economic value of water and cost recovery: Water and associated pollution management activities such as water supply, effluent disposal, drainage, debris and solid waste control, floodplain management or resource protection are often subsidized to meet social obligations. Water resource management and pollution charges and penalties should reflect the true economic value of the resource. The revenue should generate funds for management, protection, and regulation of rivers and water resources.

(vi) Emerging technologies and new management techniques: The effective management of the basin's resources requires leading-edge knowledge, technology, and management tools. Information technology and mathematical modelling and risk assessments promote the optimization of resources development and management.

(vii) Community and stakeholder participation: Consumers, service providers and the community at large should be involved in the decision-making process and be partners in ensuring the use of best practices in resource and basin management. They are also effective in promoting awareness of issues and the wise use of resources while conserving the environment.

(viii) Private sector participation: Privatization could mean right sizing of government machinery, an effective means of achieving improvements in efficiency or funding improvements in infrastructure and service. As mentioned earlier it promotes the principle of separation of regulatory and service provision functions.

KEY ISSUES

The overriding issues that had to be addressed for the Klang river basin and indeed for any river basin are:

(i) Awareness and education: Establishing the awareness of the need for river basin management and educating people of the impacts that development has on the basin, particularly environmental implications and overall economic and social costs to society if development is not sustainable;

(ii) Information management and performance monitoring: Maintaining good systems for management of information and its transfer/ sharing (including comprehensive sets of databases) and provision of skilled advice and performance monitoring, periodic reviews, and audits;

(iii) Integrated policy and strategies: Ensuring that the concept of integrated river basin management is embodied in coherent government policy and strategies including cost recovery approach, at all levels and across all disciplines and jurisdictions;

(iv) Institutional partnering and function separation: Establishing effective and transparent cross-jurisdictional institutional arrangements (both transitional and long term), separation of policy formulation and regulation from service provision, partnerships including private sector involvement, and dynamic coordination of processes and procedures;

(v) Constitution, legislation, and standards: Ensuring compliance with the Malaysian Constitution and the establishment and maintenance of legislation, regulations, memoranda of understandings, agreements, codes, standards, and guidelines;

(vi) Implementation capability, participation, and cooperation: Implementing and maintaining institutional capability, including skills, advice and training, resources and funding, participation arrangements and the willingness/cooperation by the community and all key stakeholders to achieve a sustainable outcome;

(vii) Conflict resolution and regulatory control: Ensuring that effective conflict resolution and regulatory control processes are established, and that decisions, are implemented, and non-compliance properly addressed; and

(viii) Champion profiling: Identifying a champion or champions to profile and promote the cause.

Proposals for Institutional Reforms

To deal with the problems currently being faced in the Klang river basin, a workable model established around an appropriate institutional arrangement is required. The key to determining such a model is assessment against an appropriate set of critical success factors associated with the above key issues.

As indicated earlier, there is no 'right' model for integrated management of river basins and any translation of the various international river basin management arrangements must be tailored to suit a particular country and/or region, taking into consideration actual physical and social (both political and cultural) conditions. In determining what is appropriate for management of the Klang river basin, it is essential that a set of critical success factors applicable to the basin be identified. The critical success factors for IRBM are shown in Figure 6.3.

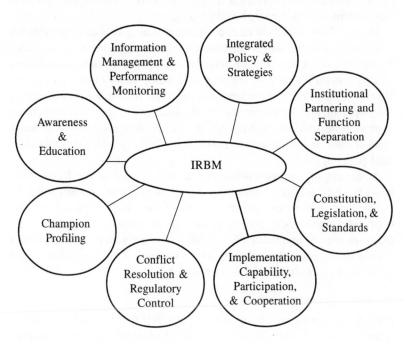

Figure 6.3: Critical factors for IRBM

At the top of the list of critical success factors is the need for an integrated and holistic approach for the basin since land development has such a direct impact on water resource and vice versa.

MODELS

The ideal model for managing the resources in a river basin will be to have a single authority vested with all the right and might to deal with river management in a comprehensive and integrated manner. It can serve as a one-stop agency responsible for overall planning, coordinating, monitoring, regulating, licensing, enforcing, and managing the river and its resources. This is best done within the ambient of a river basin master plan with inputs from all. To avoid biases and potential conflicts of interest, this agency should focus on a regulatory role and not be, at the same time, a service provider or implementer (developer).

However, in determining an appropriate institutional model for adoption, it is more prudent to first consider options that would be less disruptive to the status quo and thus more acceptable, while retaining the option to move ultimately to the most desired outcome.

In the case of the Klang river basin, there is already a set of institutions overseeing planning and development in the Klang basin, viz. the Klang Valley Planning Council, the Klang Valley Planning Working Committee, and the Federal Territory Development and Klang Valley Planning Division. However, while these institutions have a major role in land planning and development, they have limited or no involvement in other natural resource management activities, and there is only a limited consideration of water management sustainability and flood and pollution control as it relates to land planning.

In developing an appropriate institutional set-up for the Klang river basin, the easiest option would be to review the available resources within the existing institutions, ensure appropriate representation, and to extend their role within their current powers to incorporate integrated basin management. A minimum of legislative and regulatory change would appear to be required for this option to be implemented and this would be generally consistent with best practice.

PROPOSED REFORMS

With these considerations in mind, it was felt that the quickest option would be to start with changes at the level of overseeing and managing the activities within the Klang river basin and those would be centred around the operations of the existing Klang Valley Planning Council and the supporting working committee and secretariat. The Planning Council will become a management council that will not only look into planning policy, but also at management policies.

It is proposed that through an agreement between the national, state, and federal territory administrations, the Council be expanded in its coverage and renamed the Klang Basin Management Council. The working committee would become the Klang Basin Management Executive Committee, supported by an expanded FTDKVPD to be known as the Klang Basin Planning Unit.

The charter for the proposed institution would be 'To promote, plan for, and coordinate the sustainable management of the natural resources

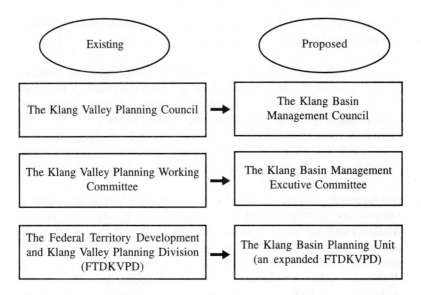

Figure 6.4: Institutional reforms proposed in Klang river basin

(water, land, and environmental) of the Klang basin'. The Council would be chaired by the Prime Minister and would be expanded to include federal ministers and executive councillors from Selangor state with responsibilities for land, planning, environment, and water resources. The mayor of Kuala Lumpur would also be a member and, when necessary, relevant economic planning ministers and units could also be involved. The Council's responsibilities would include:

(i) Providing policy direction necessary for the sustainable management of the Klang river basin;

(ii) Providing direction and overseeing the activities of the executive committee;

(iii) Approving basin perspective, strategic, and development plans, based on federal and state policies;

(iv) Evaluating the progress and implementation of all approved plans; and

(v) Reviewing and evaluating other matters which may affect the management of the basin.

The Executive Committee would be chaired by the Chief Secretary to the federal government and would consist of the chief executive officers of federal and state departments responsible for land, planning, environment, and water resources. The director general of Kuala Lumpur City Hall would also be a member. The executive committee's responsibilities would include:

(i)   Developing basin strategic and basin development plans, based on federal and state policies and the specific requirements of the basin;

(ii) Establishing guidelines and targets for natural resource management;

(iii) Initiating studies and preparing overall basin plans

(iv) Overseeing and monitoring the progress and implementation of all approved plans and activities; and

(v)   Proposing the implementation of any other matters which may affect the management of the basin.

The current Federal Territory Development and Klang Valley Planning Division (FTDKVPD) would be expanded and enhanced to become the Klang Basin Planning Unit. It would remain within the Prime Minister's department but would become an 'independent unit', responsible to the Executive Committee and Council but acting independently of any other ministries and/or government department. Its delegation of powers and authority would be equivalent to that delegated to the economic planning unit (responsible for preparing the national five-year development plans).

Its staff would include experts not only on planning and development but also on water resources, environment, and land management. This staffing profile would enable the unit to function both as a technical group in support of (the Executive Committee) and act as the Secretariat to the Executive Committee.

The various federal and state departments and agencies, local authorities, and Kuala Lumpur City Hall, as well as the district offices of the national agencies would continue as the service providers for the implementation of overall policies, undertaking, among others:

(i)   Monitoring, compliance, and enforcement of structure plans and local plans;

(ii) Flood mitigation and drainage;

(iii) Environmental assessment and monitoring;

(iv) Land control; and

(v)   Management of solid waste.

The application of the recommendations emanating from the council and Executive Committee would depend to a very large extent on the support of the service providers, particularly the local authorities where so much development/planning power resides. Hence, if this option is found to be not effective, the creation of a single 'management authority' will have to be considered. Essentially, the same organizational structure will be in place but the name and method of incorporation will be significantly changed.

Legislation will have to be used within the national arena (using the federal powers across state borders) to establish a Klang basin management authority. This authority will then have the directive and coercive powers of a government department, and its directives and recommendations for actions on plans, activities, and applications will be mandatory as it will be given powers to impose sanctions on any authority, state or local, which does not conform to the overall plans of management for the basin. This option is undoubtedly the 'hard option' and must be seen as the last resort, given the social and political implications.

COMMUNITY ADVISORY COMMITTEES (CACs)

Water problems manifest itself at the local level and, ideally, decision making regarding solutions should be carried out at the lowest appropriate level, involving all stakeholders (all levels, all sectors, public-private-community involvement). 'Best practices', globally, is showing a trend for greater degrees of civil governance, with a focus on people as the key element in the decision-making process, striking a balance between dependence on government and the technical experts (top-down) and empowerment of the community (bottom-up).

With this in mind, a recommendation has been made to the government to incorporate 'community consultation' activities within the proposed IRBM approach, centred on community advisory committees (CAC's).

The principal objective of a CAC is to assist the local authority in the development and implementation of a natural resources management plan for sub-catchment areas under its jurisdiction. The CAC would be both the focus and a forum for the discussion of technical, social, economic, and ecological issues and for the distillation of possibly differing viewpoints on these issues. The CAC would achieve this by ensuring that the community and all stakeholders (often with competing desires) are equally represented. As such, the composition and roles of CAC members are matters of key importance.

It has been proposed that the CAC would be formed and chaired by the local authority. Since, by law, the responsibility for planning matters lies with the local authority, the CAC, which is advisory in nature, should then report directly to the local authority.

Membership of the CAC would consist of a balanced mix of community representatives, councillors, administrative staff, together with technical experts from relevant agencies, who are committed to and actively involved in the preparation and implementation of a local plan for natural resources. Community representatives would include people from affected residential and business areas, together with people who can effectively inform the affected community of the deliberations of the CAC and so foster a wider understanding of the natural resource management process. As such, the community representatives play an important role in the success of the CAC and every attempt will have to be made to have representatives who are enthusiastic, energetic, and likely to 'see the distance' up to the final implementation of the plan.

In certain circumstances it may be desirable to establish a CAC involving a number of adjoining local authorities, especially when measures of structural, land use or flood response, and preparedness in one local authority area are likely to influence the effectiveness of measures or natural resources in another local authority area. This cooperative approach will take into consideration the fact that local government boundaries rarely follow catchment boundaries. Regional or catchment cooperation can result in a more holistic appraisal of issues, successful implementation of management strategies, and a more efficient use of technical expertise.

COMMUNITY CONSULTATION PROCESS

The local community has a key role to play in the development, implementation, and success of local plans. If such local plans are to be accepted and successful, it is essential that clear and concise communications flow between the CAC and the community so that affected individuals and community groups can 'have their say' and learn of their roles and responsibilities, i.e., be aware and educated.

Guidelines are being prepared to assist in this process. It is envisaged that the local authority would arrange to:

(i) Involve and inform the community, through media releases, newsletters, and public meetings, of the role and responsibilities of the CAC in the development of a local plan for natural resources. Also, affected residents should be informed of the length of time expected to elapse until finalization of the local plan and implementation of works, and of the nature of controls, pending completion of the local plan;

(ii) Define clear goals for the development of each local plan and estimate the time to complete each investigation and when direct consultation and feedback with the community is proposed;

(iii) Call for representatives of the general community and from action groups to self-nominate for the CAC, clearly stating the expected role of CAC members at this time;

(iv) Use established local community groups where they exist, and encourage representation of these groups on the CAC;

(v) Make one or two contact people known to the community, usually staff members of the local authority, who can be contacted regarding questions relating to natural resource management, both during the development and implementation of the local plan, as well as after the local plan has been adopted;

(vi) Release information to the community and members of the CAC at regular intervals, rather than waiting until the completion of one of the formal stages of the local plans, or associated formal meetings of the CAC;

(vii) Ensure that simple and clear messages are used when disseminating information, to explain the situation in uncomplicated language and relate any implications to property owners and potential building and development applicants.

The development of local plans is neither short nor simple, nor is it the singular responsibility of the local authority officers, consultants, or government officers to have input to the process. The CAC must comprise members who are committed to and actively involved in the preparation and implementation of the local plan. Of necessity, the adopted local plan will be a compromise involving trade-offs where certain individuals may be disadvantaged, others advantaged, but the community as a whole will be better off. An important role of the CAC will be to assist in the presentation and resolution of conflicting desires and requirements of various community groups and individuals.

Conclusion

Malaysia is rich in water resources and as such water has traditionally been used as an easily available resource. However, the rapid pace of development over the past two decades has resulted in water-related problems (floods, water shortages, deterioration of river water quality) becoming more frequent and acute. This has led to concerted efforts among water professionals to push for the introduction of integrated water resources management. As IWRM is not easily understood by decision makers, efforts in recent years has been to push for the introduction of integrated river basin management which is to integrate the planning and development of water-related projects in a river basin.

The Klang river basin is the premier river basin in the country and the capital city of Kuala Lumpur is located in the basin. The river itself is an interstate river in that it starts in the state of Selangor, enters the Federal Territory of Kuala Lumpur, and then re-enters the state of Selangor. Given that land and water are state responsibilities, and that the federal government has the expertise and funding, the experience gained from this case will be a useful pilot to be replicated in the other states.

The need for integrated management of resources within the Klang river basin has been propagated for quite a number of years, but there has not been any significant change in managing resources within the basin. Much of the management and action has been sectoral in nature with limited cooperation and coordination. However, as issues concerning water and rivers such as water quality, flooding, and pollution become more significant in the social and political arena, the pressure for change will increase.

A key element in IRBM is the need for an effective institutional set-up to plan and manage the resources in a sustainable manner. While a single-authority model will provide the best way to implement change, it is sometimes more politically and socially expedient to move in a less disruptive way, leveraging on what is existing while keeping the final desired model as a last option. This has been the approach followed for the Klang river basin.

References

Abdullah, K., 'Everyone Lives Downstream; The Need for Sustainable Use and Management', Malacca: World Day for Water Seminar, 1999.

Abdullah, K., and Azuhan Mohamed, 'Water—A Situation Appraisal and Possible Actions at the Community Level', Seminar on 'Local Communities and the Environment II', Petaling Jaya: Environmental Protection Society, Malaysia, 1998.

Abdullah, S., 'Towards a Malaysian and Global Vision for Water, Life and Environment', Shah Alam: Workshop on Sustainable Management of Water Resources in Malaysia, Review of Practical Options, 1999.

'Country Paper for Malaysia', Presented at the Ministerial Roundtable Dialogue on Water Sector Challenges, Policies and Institutional Development in Asia, Bangkok, Thailand, Government of Malaysia (GOM), 2002.

'Malaysia Environmental Quality Report 2000'. Department of Environment (DOE), 2001.

'Malaysia Water Industry Report 97/98,' Water Supply Branch, Public Works Department (PWD), 1998.

'National Water Resources Study, Malaysia, Volume 1–18. Government of Malaysia (GOM), 1982.

'Seventh Malaysia Plan 1996–2000'. Economic Planning Unit (EPU), Prime Minister's Department, 1996.

'Sewerage Services Report 1994–1997', Sewerage Services Department (SSD), 1998.

'The Klang River Basin Environmental Improvement and Flood Mitigation Project', Department of Irrigation and Drainage (DID), 2003.

7

INTEGRATED WATER RESOURCES MANAGEMENT IN INDONESIA

Kimio Takeya, Mitsuo Miura and Ryo Matsumaru

Introduction

Indonesia encompasses about 13,000 islands, stretching 5120 km from east to west and 1750 km from north to south. The total area of the country is 1,905,000 km^2 (Figure 7.1). Almost all regions in the nation are tropical. The average temperature and relative humidity are fairly constant throughout the year, ranging from 23°C to 28°C and 70 to 90 per cent, respectively. The rainfall variation throughout the nation is high, however. Western Sumatra, Java, Bali, Kalimantan, Sulawesi, and Irian Jaya are the regions with annual rainfall measuring more than 2000 mm. On the other hand, the islands closest to Australia, including the eastern tip of Java, tend to be dry. Annual rainfall in some areas is less than 1000 mm (Figure 7.2).

Administratively, Indonesia is composed of 27 provinces, 235 districts, 3841 sub-districts, 55 municipalities, 35 administrative cities, and 16 administrative municipalities. The total population of the country is 204 million. Java Island, where the capital Jakarta is located, is the most densely populated island, where 100 million people live in an area of 140,000 km^2.

Water Resources Management in Indonesia

Indonesia is now in the process of decentralization after the proclamation of the Law No. 22/1999, entitled 'Local Government', and the Law No. 25/1999, entitled 'Fiscal Balance Between Central Government and Regions'. Article I of Law No. 22/1999 states that decentralization is the transfer of authority of the government by the Central government to the autonomous regions in the framework of the unitary state of the Republic of Indonesia.

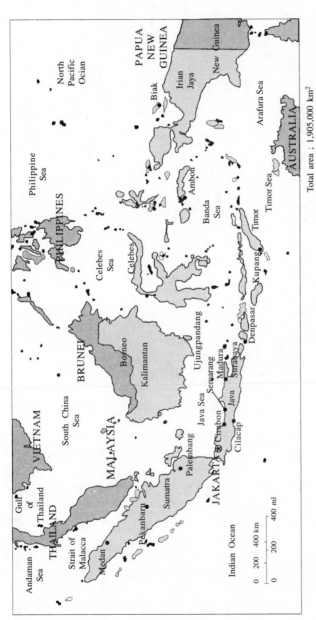

Figure 7.1: Republic of Indonesia

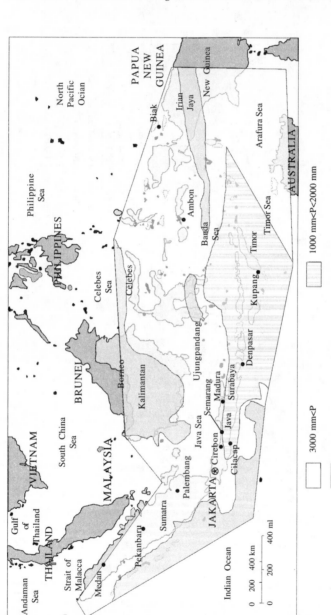

Figure 7.2: Annual rainfall in the region (mm/year)

On the other hand, regional autonomy means the authority of the autonomous region to regulate and govern the interest of the local people according to their own initiative based on the aspirations of the people in accordance with rules and regulations.

In the water sector, the present Basic Law No. 11 of 1974 on water resources (UU 11/74) and all its regulations are under the reform process in line with the above mentioned policy of decentralization.

The principles of the reform process are as follows:

(i) The role of Central government would be limited to that of an enabling and regulatory one.

(ii) Public-private partnership will be promoted at the regional and local levels.

(iii) Resources will be transferred to the local and regional levels.

(iv) Implementation authority would also be devolved to provincial, district, and local governments.

(v) Public consultations and stakeholder participation will be encouraged by creating institutions which would facilitate such dialogues and inclusion.

(vi) A participatory irrigation management system will be put in place so that responsibility of irrigation management is transferred to water user groups.

A new Water Resources Bill based on the above mentioned policy is under preparation. According to the draft of 'Indonesian Water Resources Bill', Article 7 provides for the creation of certain councils to manage river basin and water resources. Interstate river basins will be managed by the relevant ministry in consultation with the water resources national council of the concerned states. River basins that lie across provinces will be managed by the relevant ministry in consultation with the water resources rational council of the state. For river basins which overlap for districts and towns, the managing authority is the governor in consultation with the water resources provincial council. In case of river basins lying within a town or a district, the mayor or regent is the managing authority in consultation with the water resources district council.

The bill provides for the participation of the community by giving it an extensive role in water resources management (Article 5.3), recognizing the traditional rights of communities to water (Article 9.4), and providing space for community participation through a series of articles.

The World Bank extended a loan of US$ 300 million as a 'water resources sector adjustment loan', (WATSAL) in April 1999 for the restructuring the water sector of Indonesia. Under WATSAL, new institutions are proposed for the reform of the water sector. These institutions, their functions and membership are given in Table 7.1

The four-tiered structure as given in Table 7.1 will provide the framework for:

(i) The strategic river basin corporation of Jeneberang, Bengawan Solo, Juratunselna, and Serayu-Bogowonto. These will be self-financing basin water management corporations. However, membership and accountability links are not clearly spelt out.

(ii) A provincial river basin management unit (Balai PSDA) in each of the eight provinces. The unit will implement all regulatory arrangements for water allocation, discharge of wastewater, drought management, conjunctive use of groundwater and surface water, monitoring water quality, and integrated watershed management.

(iii) Provincial and river basin water coordination councils—PTAP and PPTPA. These councils will serve as multi-stakeholder institutions to develop basin-wide development plans and policies.

(iv) A national hydrology management system which will be responsible for improved hydrological data collection, processing, and information retrieval for surface and groundwater.

(v) Provincial hydrological units (PHUs) in eight provinces which will manage collection, processing, and dissemination of hydrological data pertaining to groundwater and surface water.

(vi) A national water quality sampling organization or network. This will provide cost-effective water quality management interventions by allocating financial contributions to polluters and beneficiaries of wastewater treatment.

(vii) Water users' associations comprising of rural and urban water users which will operate and manage irrigation networks. The associations will also manage large irrigation schemes jointly with the government and have autonomous governance and financial control.

(viii) Federations of water users' associations which will have representatives from the associations. The federations will be responsible for managing large irrigation schemes jointly with the government.

The World Bank is also supporting basin water resources planning (BWRP) through a grant from The Netherlands. BWRP developed a methodology for basin water resources planning, in the form of guidelines and the BWRP decision support system. The methodology was tested on three pilot basins, namely Citarum river basin, Ciujung-Ciliwung river basin, and Juratunselna Basin.

Table 7.1: Institutions proposed for the reform of the water sector in Indonesia

Name of Institution	Function	Membership	Accountability links
Inter-ministerial coordination body (*Tim Koordinasi*)	• strategic planning • inter-agency coordination • promotion of public, community, NGO participation in basin management institutions.	• Chairperson: Coordinating minister for economic affairs • Members: nine ministers	Report to the President
WATSAL steering committee	• Will function as the technical secretariat to *Tim Koordinasi*.	• Chairperson: Deputy chairman of BAPENAS • Director Generals of eleven central government ministers • Rep. from local governments (provinces) • Academia and NGOs	Report to *Tim Koordinasi*
WATSAL Task Force		• Chairperson: Chief, BAPENAS Bureau of Water Resources • central government officials • local government (provinces and districts) officials • academia and civil society	Report to WATSAL steering committee
WATSAL Technical Working Groups	• Working group on national regulation pertaining to water sector • Working group on river basin and BMCs • Working group on water quality and pollution control • Working group on irrigation		Report to WATSAL Task Force

Japan International Cooperation Agency (JICA) is supporting the integrated river basin management (IRBM) of the Brantas River through 'The Study on Comprehensive Management Plan for the Water Resources of the Brantas River Basin'.

An Example of Water Resources Management (Soil Erosion Control and Reforestation—Citarum River Basin in Central Java)

INTRODUCTION

The Citarum river originates in Bandung and flows to the north up to the Java Sea. The catchment area is 6000 km^2 and its total length is 350 km (Figure 7.3). It is one of the rivers in Indonesia where integrated water resources development have been carried out extensively. In 1975, Curung dam and Jatiluhur dam were built for the purpose of irrigation of 280,000 ha of paddy field and to provide water supply for Jakarta. In 1985, Saguling dam was built for the purpose of hydropower generation. In 1988, Cirata dam, for hydropower generation, was built between Jatiluhur dam and Saguling dam.

Jatiluhur Authority was established in 1974 as an integrated basin management authority. The basin authority leased major irrigation, water supply, and other infrastructure investments and recovered costs from water supply sales to cities, industries, and for hydropower.

The Upper Citarum watershed plays an important role to save and guard the low-lying area where vital facilities such as Saguling, Cirata, and Jatiluhur water reservoirs and their hydroelectric power stations like Plengan, Cikalong, Saguling, Cirata, and Jatiluhur have been constructed. Further, it also reduces the yearly flooded area of Bandung basin and sustains irrigation in the largest rice producing area of Indonesia in the northern part of West Java, while providing clean water supply to Bandung and Jakarta City and other surrounding areas.

The Upper Citarum watershed is not only important for the local needs but also for the regional and national requirements of electricity, and for its role in national food production.

Citarik sub-watershed, the most critical sub-watershed in the Upper Citarum watershed, is now undergoing reforestation (Figure 7.4). It is being developed under the 'Upland Plantation and Land Development Project at Citarik Sub-watershed'.

The most serious consequence would be the uncontrolled erosion and sedimentation of the flood area of Bandung basin and Saguling dam basin. Aside from the negative effect caused by the physical degradation

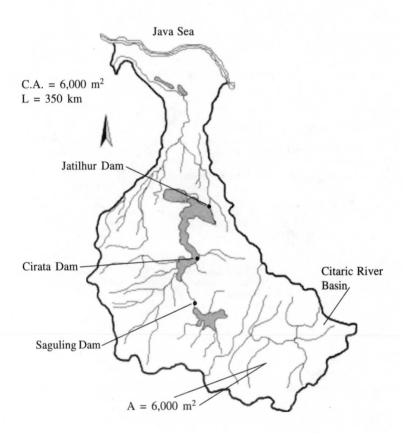

Figure 7.3: Citarum river

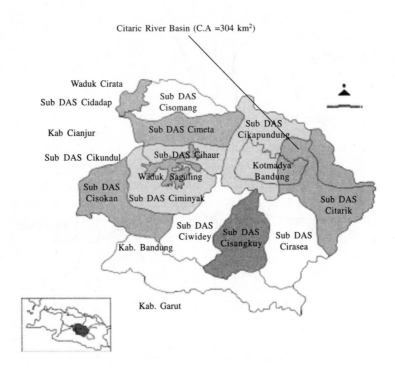

Figure 7.4: Citaric river basin

in the Upper Citarum Watershed, there is the risk of economic losses due to a reduced lifespan of the dam. Other damages over the short term could be eventual loss of stable vegetative cover, which can reduce soil fertility and water recharge. As a result of these, the downstream flat area of Bandung basin will suffer from flood and shortage of surface and groundwater, which will adversely affect many activities in various sectors, especially the life in the urban area of Bandung city. The upland area will suffer from a decrease in agricultural activities, and thus, reduce the incomes of the farmers.

OVERALL PLAN OF SOIL EROSION CONTROL

The organization in charge of the watershed management is the provincial governor, when the owner of the land is the Central government. If the owner is private, the district governor (*Bupati*) is in charge. In the case of Citaric Project, the land is owned by the private sector, and hence the district governor is implementing the project. In both cases, projects are implemented with the participation of the residents. Good watershed management requires participation of the residents, which is the key to the success of the project.

Watershed management is closely linked to agriculture development and even regional development. It cannot be sustainable unless the project benefits the local people. Accordingly, it is necessary to analyse the socio-economic reality of the basin.

IMPLEMENTATION

    (i) Projects being implemented by contract between the local government and the farmers' groups:
- community forest development
- agroforestry development
- improvement of farm management
- propagation of nursery and multipurpose huts

    (ii) Projects being implemented by contract between the village development committee and the local government with the participation of farmers:
- torrent control facilities

    (iii) Projects being implemented by contract between the contractors and the local government:
- road construction connecting villages (in this case, employment of labour and materials should be locally provided)

The unit of the project in Citarik is the smallest administration unit of Indonesia, that is a village (*desa*). The projects are carried out according to the land of each farmers' group, the area of which average approximately 25 ha.

ACTION PLAN

The draft action plan is prepared by the provincial government based upon the development action plan by the consultant. Each farmers' group make a detailed plan and promote coordination between the group and the district to formulate the final action plan. The final action plan becomes the contract document between the local government and the farmers' group. The action plan notes the names of the members of the farmers' group, activities, wages, area, crop types, fertilizer, pesticide, seeds requirements and other associated expenses.

FLOW OF FUNDS

The project fund is basically managed by the farmers' group in order to strengthen their organization and forge the spirit of self reliance and promote sustainability of the project. Thus, each farmer group opens their own account at a commercial bank and the project funds are transferred to the account by the central bank through the local accounting office.

The farmers' groups make an action plan with the assistance of NGOs and community promoters, and the fund is transferred according to the action plan. Each farmers' group withdraws funds from the bank with the permission of the community promoters and NGOs, depending upon the achievement of the group. The funds are accounted for by the accountant of the group and the account is accessible to any member of the group. Negotiations for purchasing fertilizer, pesticide, and seeds are conducted by the farmers' group with the suppliers. The community promoters and the NGOs assist the farmers' group in negotiations. The contract documents and specifications for purchasing are prepared by the local government with the assistance of consultants. Quality test of the farming equipments is done by consultants and only those that pass the test are supplied to the farmers' group. Technical training, management training, and fund monitoring activities are performed by NGOs. Monitoring of the fund and transparency of the fund removes doubt among the farmers and contribute to the solidarity of the farmers' group. The Indonesian government has been adopting a top-down method for this kind of project, where the fund flows from the top to the bottom. This leads to a large portion of the fund being diverted in bureaucratic procedures. However, in the Citarik project, a bottom-up method is being employed. This method leads to sound distribution of fund which has been planned. At the same time, conscious unity among the farmers is forged and sustainability of the project is promoted.

PROBLEMS ENCOUNTERED

The difficulty arises in combining the maintenance of quality of civil works and participation of the farmers in the project. In order to make the project sustainable, the participation of the farmers is imperative. However, lack of professional skills among the farmers cause degradation of the quality of the civil works. Some loss in quality needs to be tolerated considering the essentiality of farmers' participation.

In Indonesia, sector projects have been functioning on the initiative of the central government. The Citarik project is a multi-sector project initiated by the local government. In that sense, both the people and the local governments are facing a new regime of project implementation. There is a lot of confusion among both the local government and the farmers in this period of transition.

SUSTAINABILITY OF THE PROJECT

Regular inflow of funds is indispensable for continuation of the projects. In Citarik project, the initial fund for the first year of the project is provided by the project itself. After the harvest of the first year, the fund equivalent to the fertilizer, pesticide, and seeds is to be saved in the farmers' group for the next harvest. The methodology aims at a conscious revolution—the idea of self-reliance.

In order to achieve this, the marketing strategy of the first year is very important. Therefore, specialists for marketing (NGO) take part at the stage of drafting the action plan and select crop types and market as well as decide on processes for market development. Negotiations are entered into with the suppliers to buy a part of the harvest so that it will increase the overall income of the farmers' group.

Conclusion

Indonesia is in the process of decentralization. In this process, conflicts over natural resources, including water resources, are unavoidable. Integrated water resources management being implemented in various river basins with the support of international organizations. There is a confusion in the combination of the concept of 'integrated' water resources management and the concept of 'decentralization'. It is necessary to redefine the two terms in order to realize better water resources management in Indonesia.

References

Suwondo, K., 'Decentralization in Indonesia', INFID Annual Lobby, 2002.

Zamaan, M., 'Restructuring of the Water Sector in Indonesia: An Institutional and Legislative Challenge', World Bank Information Centre, 2002.

8

INTEGRATED WATER RESOURCES MANAGEMENT IN VIETNAM: PRESENT STATUS AND FUTURE CHALLENGES

Jan Møller Hansen and Do Hong Phan

Introduction

Vietnam witnessed a relatively high rate of economic growth in the decade from 1991 to 2000. The average annual growth rate of its gross domestic product (GDP) was 7.5 per cent, rising to 8.4 per cent during the period between the two living-standards measurement surveys in 1992–93 and 1997–98. Over a period of ten years, the poverty incidence according to national poverty standards was reduced to only two-thirds of the 1990 rate. By international poverty standards, Vietnam's poverty incidence was reduced to just half of the 1990 rate. Vietnam is considered by the international community to rank among the best performing countries in terms of poverty reduction. Nevertheless, Vietnam remains a poor country with low per capita income (per capita GDP is estimated at about US$ 400 in the year 2000) and the poverty rate is still high (Socialist Republic of Vietnam, 2002).

Vietnam has a total land area of about 33 million hectares (mha) of which 25 mha comprises mountainous and hilly regions. Land degradation in Vietnam is caused by urbanization, insecure land tenure, poor logging practices, drought, salinization, and acidification. Steep slopes and deforested landscapes, especially in the central highlands and north-west regions, are very susceptible to soil erosion during heavy rains. Salinization and acidification are more severe in the Mekong delta region. About 19 mha, i.e. 58 per cent of the country's land, is classified as forestland. Of this, only about 11.3 mha is actually covered by forests (9.7 mha of natural forests and 1.6 mha of plantations) the rest having been encroached upon by agriculture or lying bare from deforestation. Despite recent increases in area, the quality of forests remains a concern (World Bank, Danish International Development Assistance [Danida] and National Environment Agency [NEA], 2002).

Vietnam is one of the world's ten most biologically diverse countries, containing about 10 per cent of the world's species, even while it has less than one per cent of the earth's surface. The high diversity of species and its characteristic endemism is under threat from habitat losses caused by population growth, agricultural expansion, and dam and road construction. Demand from within Vietnam, and outside, fuels a major trade in wildlife.

Vietnam's rich and diverse coastal and marine ecosystems are also under threat. Over the last five decades, Vietnam has lost more than 80 per cent of its mangrove forests, shrimp farming being one of the leading causes for this destruction. The loss of mangrove forests has been largest in the Quang Ninh and Hai Phong provinces in the north-east, and the Ca Mau peninsula in the Mekong delta is also a sensitive area with mangrove forests under severe threat. About 96 per cent of Vietnam's coral reefs are severely threatened by human activities, including destructive fishing methods, excessive-fishing, and pollution.

Wetlands are among the most threatened habitats in Vietnam, with half of globally threatened birds in Vietnam being dependent on this ecosystem for their survival. However, wetlands have yet to gain official recognition as a distinct land use or conservation management category (World Bank, Danida and NEA, 2002).

Wastewater and run-off from urban areas, industrial centres, and agricultural land, pollute surface, ground, and coastal waters of Vietnam. Untreated sewerage from households, effluents from industrial enterprises, and seepage from garbage dumps or landfills are the main causes of organic pollution of surface water. It is estimated that 90 per cent of the enterprises established prior to 1995 have no wastewater treatment facilities, and use obsolete equipment.

Despite its relative abundance, water is an increasingly vulnerable resource in Vietnam. Population and economic growth compete for water to meet food requirements, domestic and industrial water consumption, hydropower, and other uses. Spatial and temporal variability of rainfall and run-off are high, even though ample water is available on average. Vietnam experiences severe floodings at certain times and water shortages at others, and watershed degradation has exacerbated these effects. There is evidence on floods in the Mekong delta and Central Vietnam, indicating increasing flooding water levels in monitoring stations during 1991–2000.

Recently, many irrigation works and hydroelectric dams have changed the flow regimes of rivers in Vietnam. The consequence of extensive irrigation is saline intrusion into groundwater in estuary areas such as Thai Binh, Hai Phong, and Quang Ninh provinces in the north, and the Mekong delta region in the south. This salinization not only affects the quality of drinking and industrial water, but also threatens ecosystems and agricultural systems. The severity of the salinization depends on the topography and flow, with areas such as the Mekong delta affected more than the Red river delta (World Bank, Danida and NEA, 2002).

Vietnam has ten river basins or sub-basins (See Figure 8.1) with catchment areas larger than 10,000 km^2 inside Vietnam and a number of smaller basins on the narrow central coast that conveniently may be grouped in river basin districts. Together, the large river basins represent 80 per cent of the country's area, 70 per cent of the water resources, more than 80 per cent of the population, and generate more than 80 per cent of the GDP of Vietnam. Economically, the most important basins are: Red river basin, Ma river basin; Ca river basin, Srepok-Sesan river basins, Dong Nai river basin, and Mekong delta.

International relations in six of the sub-basins are significant. The Red, Mekong, Ca, and Ma rivers receive a significant part of their available resource from upstream countries, while the Ky Chung–Bang Giang river discharges into China and the Srepok–Sesan rivers into the Mekong river through Cambodia. On the Mekong river, international cooperation was established as early as 1957 and strengthened through the Mekong River Agreement in 1995 (Danida and Ministry of Agriculture and Rural Department [MARD], 1999).

Trends in the country's main river basins indicate good upstream water quality, while downstream sections are often polluted. Organic pollution gets worse in the dry season when rivers flows are reduced. Lakes, streams, and canals increasingly serve as sinks for domestic sewage and industrial wastes within cities. In general, the groundwater quality is considered good, even though certain areas are affected by salinization, acidification, and point-source pollution. However, groundwater contamination is assumed as being an increasingly serious problem, especially in and around large cities and industrial sites.

Rapidly rising demands for water from rural and urban areas for agriculture, household, and industrial uses, pose a formidable challenge. The demand for water continues to escalate. Agricultural lands, (accounting for 90 per cent of water use) continue to expand. The rapid development of industry and service sectors, and a rapid expansion of large cities and urban areas

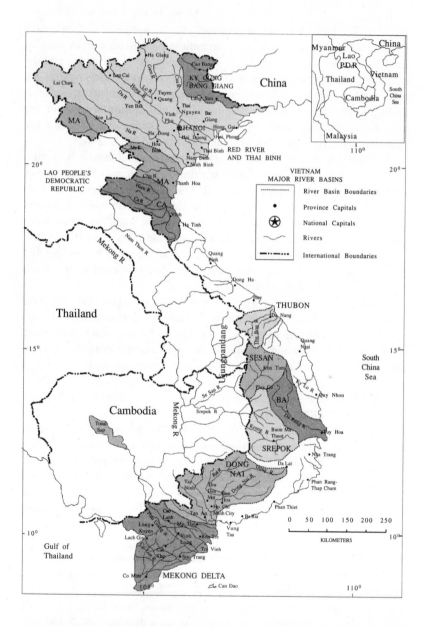

Figure 8.1: Major river basins in Vietnam

also create an increasing high demand for water supply. In urban areas, the demand for water is expected to double over the next twenty years.

Water quality is decreasing as agricultural, industrial, and household users vie for scarce water supplies, and the development of effective rural domestic water supplies and sanitation has only been initiated over the last 5–7 years. Rapid urbanization and industrialization in the interior, development port and marine transport, expansion in coastal tourism, and rise of oil spills contribute to the deterioration of the quality of coastal water (World Bank, Danida and NEA, 2002).

To address these challenges, the government will need to accelerate reforms in the central and local administration in general, as well as in institutions dealing with water resources and environment, and create and enforce an effective framework for the water resources and environment sector that will allow the private sector, NGOs, and civil society to respond and take full part in the development process.

Integrated water resources management (IWRM) can be defined as a process which promotes the coordinated development and management of water, land, and related resources, in order to maximize the resultant economic and social welfare in a equitable manner without compromising the sustainability of vital ecosystems (Global Water Partnership, 2000).

This chapter describes briefly the overall status and some main challenges of putting IWRM into practice for securing sustainable water resources and environmental management in Vietnam.

Water Resources Sector Framework

Vietnam has put in place a sound legal framework for water resources management, environmental protection, and natural resources conservation, which includes many laws, regulations, and directives.

The Law on Water Resources (LWR), which provides for water resources management reforms, was passed by the National Assembly in May 1998. The law took more than ten years to develop, and its formulation was assisted by the World Bank. A number of donors are supporting the implementation of the LWR, including the Asian Development Bank (ADB), World Bank, Danida, Australian Aid (AusAid), Netherlands, International Union for Conservation of Nature (IUCN) and other donors.

Other important legislations affecting water resources management in Vietnam include, among others, the following: i) the Land Law, which was enacted in 1993 and deals with the ownership of land, land use rights, and government regulation of land use; ii) the Law on Environment, which was enacted in 1993 and broadly defines the environment, recognizes the 'polluter pays principle', use of appropriate and cleaner technologies, reduction in wastewater discharges, solid waste management, requirements for environmental impact assessment, standards and inspections and their enforcement; and iii) the Forestry Protection Law, which defines the so-called state management of forest and forestry lands, regulation of forest use, and rights and obligations of forest users.

An official orientation, a draft national strategy and a number of decrees exist for urban water supply and sanitation, which are enforced by the Ministry of Construction (MoC) and its provincial departments.

A National Rural Water Supply and Sanitation (NRWSS) strategy was approved by the Prime Minister in August 2001, which promotes an approach of rural water supply and sanitation based on demand-responsiveness, informed choices and affordability, community and user-based operation and maintenance systems, and participatory approaches for planning, implementation, and management of rural water supply and sanitation (MARD and MoC, 1999). The NRWSS strategy has been developed with the assistance of Danida, and a number of donors, including Danida, AusAid, Unicef, World Bank and other donors, are assisting the government for implementing principles as mentioned in the NWRSS.

In recent years, national and international workshops and seminars in water resources management have been organized in Vietnam to boost and enhance advocacy and exchange of international knowledge, expertise, and practice within water resources management. For example, MARD, in association with Global Water Partnership and leading donors, organized a national IWRM workshop in October 2001, and other workshops were organized in late 2002 on thematic issues such as water, food and environment, and other important issues related to IWRM (MARD et al., 2001). There is a growing awareness and understanding of the importance of and the need for IWRM, and also a keen national interest and willingness to learn and explore from experiences and new initiatives in the region and from other parts of the world.

The LWR and NRWSS strategy place the major responsibility for attaining the government's water vision with MARD. MARD also plays a central role in the support to the National Water Resources Council (NWRC) and establishment of river basin organizations (RBOs)[1]. It should

---

[1] In Vietnam, the term river basin planning organization is recognized by the government in order to emphasize that these organizations are advisory bodies only and not executive or management entities with decision-making power. The term river basin organization is, however, used in this chapter for purpose of clarity.

however be mentioned that following the election for the National Assembly in May 2002, the Government of Vietnam has established a new Ministry for Natural Resources and Environment (MONRE). The establishment of this new Ministry has created some doubt in the central administration and among donors regarding the future national mandate, responsibility, and anchorage of water resources management in Vietnam.[2]

It is estimated that there are at least ten ministries, fifteen central committees and general departments, and many more water and environment research and consultancy institutes and centres, universities and other institutions that influence and operate in the water resources and environment sectors. The ministries, departments, institutes, and other agencies are highly segmented with limited cooperation among them, and experiences show that these institutions often have inadequate capacity and suffer from over-staffing.

A number of initiatives have been taken in recent years to enhance coordination and collaboration within the water resources sector in Vietnam. In 2001, MARD took steps to enhance a government-led coordination of donor assistance to the water sector with the establishment of the so-called Technical Advisory Group 2 (TAG.2) under the umbrella set-up known as the International Support Group (ISG). The TAG.2 meets regularly with representation of MARD and all major donors, and has proven to be a good forum for coordination between MARD and donors within the sector. However, there are still potentials for developing this forum to become a more effective mechanism for policy and strategy dialogue between the government and donors. Initiatives have been taken by TAG.2 to support the development of a national water resources profile and a review of ODA assistance to water resources management in Vietnam. Today, involvement in the water resources and environment sectors are compartmentalized and administratively confined to an extent that achieving effective and integrated water resources and environmental management is a long-term goal. In realizing the water vision of Vietnam, the government will need to enhance the coordination and collaboration between various ministries and central departments and institutes, local authorities and other actors in the sectors related to water and environment as well as to enhance the involvement of other partners in water-related sectors.

[2] MONRE is responsible for water resources management from 2003. However, not all of the tasks, as indicated in LWR, have been transferred from MARD to MONRE. The transition will take some more time.

The Law on Water Resources (LWR)

The LWR became effective in January 1999 and represents a first important step towards attaining the vision for IWRM in Vietnam.[3] The LWR and its main implementing decree define the scope of water resources management in Vietnam and also identify the responsibility of each agency at national and provincial level in the implementation of its articles. An important principle embodied in the LWR is that the ownership of water rests with the people of Vietnam and is to be managed by the government on their behalf. Article 58 of the LWR describes the responsibility of government for the management of water resources. It specifies, among other things, that the government, through MARD, performs the role of water resource manager, with comprehensive and far-reaching powers. Other ministries are assigned responsibilities in implementing specific functions of water resource management and water services delivery. The people's committees of the provinces and cities are responsible for management of water resources in their own jurisdiction (World Bank, 2000).

The LWR has created the so-called specialized inspection of water resources, comprehensively legislated by Articles 66 and 67, which allows for employment of inspectors who are necessary for operation and enforcement of licensing. However, the government still has to institutionalize the provision of inspection and allocate resources for its implementation and function. Often, countries have carefully drafted policies, but with very poor enforcement and operational powers. Water administrations are, as a result, quite ineffective in practice (Solanes, 2001). Though provisional powers and mandates have been described and endorsed, in reality such provisions still have to be put into practice by the government.

The LWR establishes the river basin as the primary unit of planning and management. Article 59 specifies that the National Assembly will decide on investment strategies for water resources works of national importance. The government will approve planning of large river basins and important water projects, while MARD is responsible for approving the planning of river basins and hydraulic works systems, under delegation of the government. The LWR, however, does not clarify that river basin planning concerns management as well as development.

---

[3] After five years of implementation (1999–2003), the LWR is being reviewed and modified to reflect the new developments in the water sector.

In Article 63, the LWR stipulates that the major role of the NWRC is to advise the government on important water resources issues and to coordinate national water resources planning and management across the various ministries. One of the initial key tasks of the council will be to commission and oversee a national water resources strategy and action plan. The national water resources strategy and action plan have not yet been worked out, but some first initiatives have been taken for its preparation.

Article 64 establishes the concept of a river basin organization as the vehicle for the management of river basins. RBOs are to be established under MARD. The government is giving priority to RBOs in major river basins, including the Red river, Dong Nai, and Lower Cuu Long river basins. Some decrees have been endorsed for the establishment of RBOs in these river basins, but in reality much needs to be done for setting up operational RBOs. Other important provisions of the LWR include the introduction of licensing for surface water extraction and a permit system for wastewater discharge. A preliminary decision on regulation of groundwater, including drilling and granting of permits for groundwater protection and exploitation, was introduced in March 1997 (VNMC et al., 1999).

The decision on groundwater management from 1997 has not been widely implemented and new legislative work and pilot activities on groundwater management, exploitation, and protection were initiated by MARD with the assistance by Danida in 2001. Today, there is no effective or workable system in Vietnam to control and monitor wastewater discharges from industries and domestic users. With the present economic and industrial development in Vietnam there is no doubt that this will be a major issue in the years to come if minimal environmental standards are to be maintained in the country.

The successful implementation of the LWR and initiatives for participatory management of water services will require major reforms and changes in the political and administrative systems at central, provincial, and local levels. The current institutional arrangements for water resources management reflect a vertically oriented and fragmented sub-sectoral approach, which does not facilitate full coordination and collaboration among various ministries, departments, and provincial authorities. MARD has, however, taken some constructive steps in enhancing the national coordination and exchange of knowledge and information among partners within water resources management. It is hoped that horizontal collaboration among ministries at the national level and among provinces at the river-basin level will be introduced over time. The establishment of the NWRC and RBOs are seen as important elements in the process for working towards a more integrated and coordinated approach to water resources management in Vietnam.

The government and MARD have been successful in attracting substantial donor assistance from the World Bank, ADB, Japan International Cooperation Agency (JICA), Danida, AusAid, The Netherlands, and other donors for support to water resources management. However, the government has yet to provide the required national finance and human resources necessary to fully implement the LWR. Institutional and public administrative reforms of MARD and other ministries have to be introduced in order to develop the role and enhance the capacity of concerned ministries to implement the LWR effectively. Initiatives have been taken by the government, with the assistance of UNDP, for public administrative reforms, including reforms in MARD. It is still too early to conclude on the results of these initiatives.

The limited capacity and capabilities to implement the new approaches to water resources management are somehow recognized by MARD and by the departments of agriculture and rural development (DARD) at provincial levels. However, the changes will take considerable time. Retraining and reorientation of institutions and technical staff with skills in irrigation, drainage, flood control, and water supply, to become water resource managers pose a major challenge for the government, MARD and the provincial DARDs. It is also a major challenge today to ensure a broad-based consultation and involvement of various government actors in the IWRM process, such as various ministries, departments, and institutes, across traditional sectors boundaries. Coordination and collaboration at provincial and local levels, and across provincial boundaries is another major challenge. The importance and challenge of working for a better and more effective involvement of and consultation with civil society, private enterprises, farmers, and other water users and interest groups cannot be underestimated either in the development process of Vietnam.

There are limited opportunities for obtaining a good university education in water resources or environmental management in Vietnam, and the country is faced with a great challenge as to reorient and develop its higher education system to international or regional standards. Some donors have launched development assistance in order to meet the needs for reorientation of higher education in water resources and environment. To mention a few examples, The Netherlands is providing assistance to education in coastal zone management, while Danida is focusing on reorientation of higher education in integrated water resources management and demand-responsive approaches to water and sanitation. This assistance has been under way for only a couple of years, and it is still too early to determine to which extent such assistance can impact on the reorientation and development of the present higher education system.

The higher education system focuses on traditional science and engineering knowledge and skills with limited priority being given for developing analytic and problem-solving skills and capabilities among the university students. There is no doubt that the reform and development process in the political and administrative systems need to be effectively supported by reforms for a more modern university system. In recent years, interesting initiatives have been taken in Vietnam to establish private and modern university branches affiliated with international universities, but this is mainly within business administration, human resources management, and other curricula.

It is obvious that the Vietnamese administration is in need of young graduates with new skills and capabilities in order to reorient and develop the public institutions, including the institutions dealing with water resources and environmental management and administration. It is equally important that new institutional cultures and modern human resources management systems are introduced in the administration in order to create the enabling environment that will allow for new thinking and innovation towards integrated water resources and environmental management.

Policy formulation is also a core government role in water resources and environmental management. Through its policies, government can delimit the direct and indirect activities of all stakeholder groups, including itself. It is believed that appropriate policies can support and enhance a participatory, demand-driven, and sustainable development. Policies that encourage integrated water resources management include reference to the nation's wider social and economic objectives that make up the development of the society (Global Water Partnership, 2002).

The Government of Vietnam is committed to implement the so-called 'state management' of water resources, which clearly indicates its focus on the role and involvement of state agencies and institutions in water resources management. This policy also applies to most other sectors. It is fair to say that the past and present policies and practices in general have given priority and preference to the involvement of state agencies and institutions. However, the recognition and involvement of civil society organizations and other stakeholders still remains to be sufficiently supported and developed to meet the present and future requirements for an effective water and environment sector framework. It is still a common and widely accepted view among government officials and administrators that water resources issues are to be dealt with mainly by state and government agencies and civil servants. It is however the belief of many professionals that there is an untapped potential for handing over the right and appropriate mandates, roles, and responsibilities to farmers, interest groups, and water and environment user groups in Vietnam.

It is our assessment that the LWR is a general law that does not sufficiently and clearly describe operational directives necessary for implementation of the articles included in the law. In other words, the government needs to develop, endorse, and support the implementation of appropriate strategies and action plans. It is doubtful to which extent the LWR has been effectively introduced and disseminated in the administration and among the relevant agencies and institutions at central and local government levels, who are supposed to be in charge of overseeing implementation and enforcement of the articles described in the law. It is evident that many local authorities are not aware of the existence of the LWR and its implications. It is also widely recognized that the provincial authorities function with a high degree of administrative and legislative autonomy, which give ample room for local interpretation and enforcement of the LWR. A number of other decrees on water works, urban water supply, and other water-related decrees have also been endorsed to support the LWR during the last decade. However, much more needs to be done for their enforcement as well as for development and enforcement of other additional or supplementary legislation.

Governing Bodies for Water Resources Management

NATIONAL WATER RESOURCES COUNCIL (NWRC)

The NWRC was established in June 2000 by a decision issued by the Prime Minister. The decision provides the NWRC with a clear mandate to advise government on important water resources issues and to coordinate national water resources planning and management across various line ministries (MARD, 2000). One initial key task of this Council is to commission and oversee a National Water Resources Strategy and Action Plan. Sensibly, the NWRC is advisory and not executive.

The NWRC has however met only a few times since its establishment. During the last one and a half year the Council has met only once.[4] The

---

[4] From the time MONRE took over the responsibilities for water resources management in 2003, the NWRC has met regularly twice a year and has functioned as expected.

members of the Council are high-level decision makers and officers with many duties and responsibilities, and it has to some extent been difficult to set-up and activate the Council. In a meeting of the international support group in September 2002 between MARD and the donors, a joint statement of concern was presented to MARD on the passiveness of the Council and its Secretariat. There are many examples in various countries of executive councils not performing their roles properly, thereby stalling water management (Solanes, 2001).

It is important for the Council to develop as an effective institution in order to guide and oversee the development of integrated water resources management in the country.

Ministries and Central Institutions

The water resources and environment sectors in Vietnam, in common with many other countries, suffer from fragmentation across several line ministries. The fragmentation was significantly reduced with the concentration of the most important functions within the Ministry of Water Resources in 1994. This ministry was later abolished and the water resources mandate was transferred to MARD in 1995. However, the concentration was not accompanied by sufficient consolidation and support to facilitate integrated approaches to resources management and protection.

The institutions and the revised division of responsibilities outlined in the LWR are supposed to enable a clearer division of responsibilities that potentially may continue the trend towards reduced fragmentation. The LWR assigns the responsibility for overall water resources management to MARD, making it clear that in the future the main role of other line ministries will be in water services delivery within their respective sectors. Supporting the reorganization of the water resources sector will require a combination of strong political will and flexible implementation arrangements to enforce key provisions of the LWR and to overcome entrenched sector interests. During the years, much concern has also been raised over the dual role of MARD as custodian of the water resources on behalf of all sectors and all regions, and as the provider of services within irrigation, drainage, flood control, and, most recently, also rural water supply and sanitation (Danida and MARD, 1999). It is subject to considerable debate whether the functions of water resources management shall remain within MARD or be shifted to a higher level in order to safeguard the interests of other line ministries. The establishment of the NWRC is an attempt to develop an inter-ministerial body for water resources management at national level.

The administrative reforms involved in the formation of MARD were supposed to simplify the governmental structure by combining within the institutional framework water resources management and irrigation in Vietnam. Firstly, it brought together in one ministry the most important aspects of water resources management, i.e. the responsibility for development and conservation of the water resources, flood protection and preparedness, agriculture as the main consumer of water and forestry, which plays the most prominent role in watershed protection. Secondly, the importance of water supply and sanitation in rural development was recognized by transfer of the highly fragmented responsibilities for rural water supply to the ministry. Thirdly, the need for integrated planning and management of surface and groundwater resources was recognized with the transfer of the responsibility for groundwater management from the ministry of (heavy) industry to MARD. In practice, the concentration of functions within MARD is neither complete nor fully accepted by other ministries. One reason for the continued duplication of functions is the tradition for line ministries to perform all functions within their area, and the competition for control over the government funding for implementation (Danida and MARD, 1999). Lately, new uncertainty regarding the national mandate and responsibility for water resources management has been added with the formation of the new Ministry of Natural Resources and Environment and the speculations about its future role in relation to MARD and their present portfolio in water resources management.

Table 8.1 lists the selected water sector functions based on current legislation in Vietnam.

PROFESSIONAL ASSOCIATIONS AND WATER PARTNERSHIPS

In recent years, there has been an increasing interest in forming so-called partnerships to boost a more broad-based involvement of different actors and to create cooperation across traditional administrative boundaries. Following a national workshop in October 2001, the Vietnam National Water Partnership (VNWP) was established through a decision by MARD in February 2002. The VNWP has nearly fifty different partners, among others government institutions, research and academic institutions as well as associations and a few non-governmental organizations (NGO). An organization such as the Vietnam Union of Science and Technology Association (VUSTA) has also become increasingly instrumental in enhancing a more broad-based involvement of water professionals outside the state and government administration. The VNWP has only been established recently, and the partnership is now in the process of getting fully established as a new actor in the water resources sector in Vietnam.

Table 8.1: Overall division of selected water sector functions based on current legislation, Vietnam (Modified from Danida and MARD, 1999)

Ministry	Function
Ministry of Health	Drinking water quality standards (partly practised)
	Control of drinking water quality (partly practised)
MOSTE	Monitoring of surface water quality
	Wastewater discharge quality standards (partly practised)
	Industrial wastewater discharge control (partly practised)
MARD	Drinking water quality standards (not practised)
	Surface water allocation (not practised)
	Groundwater allocation (partly practised)
	Monitoring of surface water quality (not practised)
	Rural water supply and sanitation
Ministry of Industry	Industrial groundwater allocation (partly practised)
	Monitoring of ground water quality
Ministry of Construction	Planning, design, and approval of large urban water supply
	Planning, design, and approval of large drainage, sanitation, wastewater treatment
Ministry of Natural Resources and Environment	Not yet decided
Local Government	Water resources management and service delivery in their jurisdiction

RIVER BASIN ORGANIZATIONS (RBOs)

The concept of RBO has been endorsed by the LWR in 1998. In principle, the role and function of RBOs are known to some central authorities; however, in practice much needs to be done in developing functional and active RBOs.

In recent years, a number of donors have started providing assistance for institutional and capacity building at the basin level, including the support to development of simple RBOs and related water resources management activities. For example, ADB is providing assistance in the Red River and Dong Nai basins, AusAid in the Mekong delta, and Danida is planning to assist in the Srepok and Ca river basins. JICA is providing assistance to river basin planning in fourteen different basins as preparation for specific investment projects.

The recently formed RBOs still reflect government priority being given to the exclusive involvement of government officials and civil servants, many with a technical background in irrigation. There is significant scope for developing RBOs that are composed of decision makers and representatives from various water interest groups. The distinction between decision makers and interest groups, on the one hand, and government's technical personnel, on the other hand, is still not fully recognized. Some experiences gained during the last couple of years show that a technocratic and narrow sub-sectoral approach to the concept of RBOs is hampering the development and effective functioning of such bodies.

From a legal point of view it is crucial that functions allocated to RBOs do not overlap with other organizations, that they are backed up with appropriate legal powers and capabilities, that they are organized according to the functions to be performed, and that they are ensured operational resources to effectively discharge their responsibilities (Solanes, 2001). With regards to RBOs in Vietnam it is a major challenge to develop coordination and cooperation among local governments and provincial authorities. It can be argued that effective coordination and collaboration between local governments has still to be developed in Vietnam.

Local Authorities and Civil Society

PROVINCIAL, DISTRICT, AND COMMUNE GOVERNMENTS

At the provincial level the responsibilities of MARD are entrusted with the Department for Agriculture and Rural Development (DARD),[5] which administratively belongs to the provincial people's committee, but is responsible for ensuring that regulations and technical standards issued by MARD are adhered to. Many local authorities in Vietnam are thinly staffed with limited financial and manpower resources to develop and maintain a minimum of capacity and capabilities to perform IWRM activities, and most often priorities are directed more towards infrastructure investments rather than for planning processes and monitoring requirements.

The relationship between the national and provincial levels is in theory repeated at district and commune levels, but in practice few districts and

[5] Following the transfer at the central level, the provincial Department for Natural Resources and Environment (DONRE) has replaced DARD as the provincial water resources management agency.

even fewer communes have the capacity for active water resources management. The communes and towns are the lowest formal administrative level represented by people's committees and their administrations. The commune people's committee is responsible for implementing and coordinating smaller water resources projects with the district-level administration and the advisory centre. The commune authorities carry out most of the government support functions in association with individual households and mass organizations (particularly the women's union and farmers' union), advisory centres, and banks.

The village authorities are a vital link between the formal commune people's committee and water users. Where local water users are interested in water resources activities, the village is the channel through which information is delivered to the commune people's committee. In many cases, existing community structures form a good basis for informal village user groups. The effectiveness of grassroots organizations and associations is crucial to the process of stimulating demand, and for local authorities to respond to their demand.

Figure 8.2 provides a simplified overview of the set-up of the local government system in Vietnam.

International experiences show that the role of local authorities and governments in supporting IWRM is particularly strong where there are moves towards decentralization and democratization of planning and resource management. The local authorities might in such cases offer a strong forum for local participation and can be instrumental in providing information and supporting dialogue among stakeholders and policy makers. An important role that should also be performed by the local authorities includes various measures for regulation and monitoring of water use, including economic instruments to influence the behaviour of local residents. Such instruments include rate structures and charges, fees for permits, fees for bulk abstraction of surface water and groundwater and other government services, special taxes and surcharges, which carry incentives to enhance environmentally friendly behaviour and performance as well as fines and penalties for violations. Local authorities can, through by-laws, regulate and reinforce centrally introduced water resources policies (Global Water Partnership, 2002).

In Vietnam, water management is, in practice, delegated primarily to the provincial level. In principle, provincial irrigation management companies are responsible for irrigation management from headwork to secondary cannels for larger schemes, and local communities and informal user groups are responsible for schemes from secondary cannels to tail end.

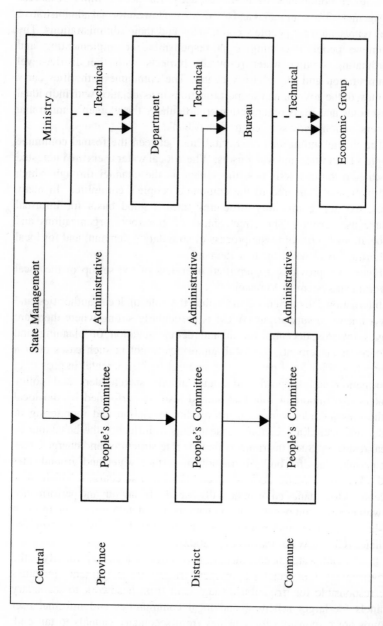

Figure 8.2: Simplified overview of the set-up of the local government system in Vietnam (VNMC et al., 1994)

However, the role and function of provincial irrigation companies are in general unclear, and reorientation of these companies will be needed in order to establish proper water rights systems and to decentralize operation and maintenance responsibilities to local communities and water user groups. The reorientation and delegation of the management of water resources and the state management of services from the provincial level to local level will require a major capacity building and strengthening effort. Legalisation of water user groups will be essential for developing sustainable management models at local levels. Developing regulatory frameworks and economic instruments for water and sanitation service delivery at provincial, district, and commune levels are also required.

WATER USER ASSOCIATIONS AND WATER USER GROUPS

The grassroot democracy decree[6] enacted by the government in May 1998 offers a legal framework for increasing community participation at the local level. It also represents an important element of an emerging enabling environment for civil society by strengthening partnership with other organizations working for the people's benefit and eventually putting in place a legal and policy framework for associations and domestic NGOs (GoV, 1998a).

The decree includes chapters on i) general provisions; ii) work to be informed to the people; iii) work to be directly discussed and decided by the people; iv) work to be discussed or consulted by the people and decided by the commune people's councils and/or people's committee; v) work to be supervised and inspected by the people; vi) building of village or hamlet communities and vii) implementation provisions.

Grassroot participation has since improved significantly in a number of communes. However, there is still ample scope for developing grassroot democracy in most communes and in society in general. There is a great scope for legalizing and actively supporting and practising the full involvement of local communities and water users as planners and managers of local irrigation and water supply schemes around the country.

---

[6] The Grass Democracy Decree states that 'Bringing into full play the people's mastery must be closely linked with the mechanism of the Party leadership, the State management and the People's mastery, the representative democracy regime must be promoted, the working quality and efficacy of the People's Councils and the People's Committees must be raised, the direct democracy regime must be well implemented in localities so that the people can directly discuss and decide important and practical issues which are closely associated with their interests.'

Some professionals would argue that the farmers and local communities always have been the real and actual practitioners and managers of water in agriculture and villages, especially when it comes to small-scale irrigation and water supply. There is however a great challenge for government, decision makers, and water advocates and professionals to promote and encourage a general and effective movement and development towards recognition of the farmer and community knowledge and expertise as well as for developing their potential mandate, roles, responsibilities, and functions in water management in Vietnam. Many development and water professionals might be of the opinion that a lot more needs to be done to recognize the role of farmers and communities in water management in Vietnam. The government, central, and local authorities, and therefore also the senior officials employed in central and local institutions, all have important roles in promoting and supporting a development towards farmer- and community-oriented water management at the local levels.

It is well known that civil society organizations are most effective in societies where there is a commitment to participation and consultation. However, in Vietnam it is commonly experienced that many local stakeholders such as local communities, farmers, and urban residents are not sufficiently connected to the decision-making processes and therefore not involved in the real dialogue with decision makers in connection with water-related investments.

There is no doubt that civil society organizations and community-based organizations can play an important role in promoting IWRM. It is widely recognized that these organizations, if allowed by government and encouraged by the enabling environment, often have considerable ability to advocate on behalf of nature and environmental protection, increase 'water literacy', that is, awareness of the need for sustainable water management, advocate on behalf of the poor and marginalized, and mobilize local communities to get involved in the management of local water resources and water delivery. Local organizations can play a strong and effective role in management of local water resources, for example for management of local aquifers and small river basins, and as planners, implementers, and managers of irrigation and rural water supply schemes (Global Water Partnership, 2002).

Water Resources Assessment and Information Management

It is widely recognized that IWRM requires a foundation of good physical and socio-economic data. Well-functioning information management systems and procedures are a prerequisite in any IWRM system, and without an efficient horizontal and vertical exchange of data and information across sectors and between administrative levels it is not possible to put IWRM into practice.

Working towards an effective exchange and management of data and information is one among several major challenges for the public institutions in the water resources and environment sectors in Vietnam. Water-related data and information are collected and stored by many different institutions, but it is often difficult to assess such data. The water resources and environment sectors are characterized as being fragmented and diversified, and traditionally there has been reluctance to share and exchange data and information. Many public institutions and government staff trade with data and information, and there are numerous examples of projects and partners who, therefore, cannot obtain and assess water data and information.

A study conducted by Danida concluded that inventory and assessment of surface water, groundwater, and water quality need to be improved. This should include inventory of water demand and discharge of waste-water in relation to licensing of abstractions and discharge, and coordinated data systems, planning, and decision tools and public information procedures are needed. With regards to administration of water rights, effective water management among others also depends on the quality of public records and registries, both of water supplies and of rights. Therefore, regulations on records and registries should be a priority. Without such records, rights are difficult to prove and certify, and their legal certainty compromised. In turn, it is impossible to have good water records and registries, without a highly effective and competent institutionalization of water management. As systems are being developed for water licensing and permits in Vietnam, it is important that the government allocates the financial and human resources necessary to implement and maintain such monitoring systems (Solanes, 2001). There are a number of donor-assisted initiatives directed towards improved data handling and management in the water resources sector in Vietnam. For example, AusAid is providing support for general water resources data management, and Danida provides assistance for development and establishment of a licensing system for groundwater abstraction.

It is a challenge for the Vietnamese authorities to improve data collection and storage methods, systematizing analysis and ensuring data quality and accuracy, sharing data and information in water resources and environmental management. Without certain requirements and a strong foundation of data and information it is difficult to develop a system for effective decision making for sustainable and financially-viable policies, strategies, and investments. It is, however, often the case that relatively low priority is given by the government for support and assistance to data collection, management, and monitoring systems to ensure a basis for broad consensus decision making and prioritization. It is still common practice that financial allocations are mainly directed towards various types of water infrastructure rather than towards more qualitative initiatives. At present there seems to be limited incentives for building up necessary water-related data collection and monitoring systems.

Provision of Water Services

Water supply companies at the provincial, city, and town levels are primarily responsible for the delivery, operation, and management of water supply. Similarly, environment companies are responsible for urban environment, drainage, solid waste management and, in recent years, also for sewerage and treatment of wastewater. Urban water supply and sanitation has during the 1990s seen a major upturn with increased government investment supported by the World Bank, ADB, Danida, AusAid, and other bilateral donors. But sector performance continues to be hampered by institutional, human resources, financial, and technical constraints such as:

   (i) urban water and sanitation companies not having clear definitions of their roles and responsibilities, nor a clear specification of their rights and obligations in relation to customers;
  (ii) the dual role of the provincial departments of construction (DoC) as the implementing agency and responsible body for approval of technical planning, design, and costing of construction projects on behalf of MoC leading to undue interference in all stages of project implementation and causing delays in project implementation;
 (iii) private sector participation being hindered by competition from sector institutions who operate both as public service providers and as commercial operators with no clear distinction between the two;
  (iv) management staff being short of skills to implement a cost-effective and consumer-oriented service delivery;
   (v) a shortage of technical staff capable of operating modern water and wastewater treatment plants;
  (vi) low staff motivation due to lack of autonomy to apply adequate incentives and sanctions;
 (vii) current investment levels which are well below estimated requirements and will take time to raise;
(viii) doubtful financial sustainability due to lack of a clear policy on cost recovery;

(ix) automation equipment not necessarily available locally and needing to be imported; and

(x) appropriate and affordable technologies for wastewater treatment need to be developed (Danida and MARD, 2000).

Many water supply companies operate with relatively low performance in terms of limited financial recovery, high percentage of water loss and leakage, and sometimes also with inappropriate technical solutions and expensive schemes in terms of high capital investment and high operation and maintenance costs. There is a tendency to focus on expansion of networks and supply capacity rather than on minimizing water wastage, and making sure that households and larger consumers are being connected to the schemes. It is a common practice that water companies operate on a supply-oriented philosophy rather than on demand management. Much remains to be done in restructuring the water supply and environment companies, and to develop a well-functioning system of regulation and enforcement of these companies in order to ensure high-quality service delivery with emphasis on consumer satisfaction as well as financial and institutional viability (Hansen, 2001).

The introduction of proper water supply tariffs and fees is a relatively new phenomenon in Vietnam. The present government decree allows only for partial cost recovery of operation, maintenance, and depreciation. In recent years, it has also become increasingly important to introduce user payment for wastewater discharge. At present the provincial authorities can only levy a wastewater charge at maximum 10 per cent of the water supply fee. This is done only in a few cities in the country, and in all such cases these are estimated as being far below a level of user payment necessary to ensure viable and sustainable sewerage and wastewater treatment services. The government and MoC have for some time been working on a new decree that should make it possible to introduce new tariffs in water supply, drainage, and sanitation to secure a more partial cost recovery. It is a major challenge for the Vietnamese authorities to ensure that financial instruments become an important part of water services delivery, and hence resources management, in the country.

The water supply and environment companies are state institutions which function largely as provincial departments with limited autonomy in business and financial planning and management. Most companies depend on subsidies from the provincial or central authorities for operation and on their approval for all levels of investment, thus constraining both initiative and incentive for efficient consumer-oriented management. Much, therefore, remains to be done for restructuring the water supply and environment companies into more autonomous and business-oriented public service enterprises. It will also be necessary to develop performance indicators enabling objective evaluations of the performance of such public agencies. There is a policy commitment with the government for developing more autonomous and business-like urban utilities or public service enterprises. However, the development and reorientation of these institutions are in general slow and not always sufficiently supported by the relevant central and local authorities.

Another important issue is the division and relations between state or public service providers and the private sector. In many cases the distinction between public institutions and private enterprises is not obvious in Vietnam. Many water supply and environment companies act as implementers of functions, which ought to be outsourced in order to enhance quality and ensure appropriate pricing of services to customers. There is also a great need to develop the involvement of small-scale private service providers and contractors to operate in the water supply and sanitation sector in order to meet the needs and requirements of customers and users.

With regards to irrigation, various experiences have shown that Vietnam's many irrigation and environment companies and institutions have considerable experience and capacity to design and execute infrastructure projects and manage complex irrigation and drainage investments. However, many lessons have been gained in recent years, which clearly indicate that the past focus on primary infrastructure has to be shifted towards a more participatory and holistic approach to design, implementation, and management in order to achieve sustainability of such investments. Many past infrastructure projects in irrigation, drainage, water supply, and other types of water-related infrastructure projects were planned and implemented without the necessary involvement of farmers, water users, and communities (ADB, 2002). There has been, and still is, a strong focus on infrastructure investments often without the necessary consultation and involvement of farmers, communities, and water user groups. This phenomenon is however also experienced in other countries, and is not necessary unique for Vietnam.

Water-related infrastructure investments are often undertaken without the necessary commitment and agreements from the project actors to ensure that administrative and institutional reforms and developments take place prior, during, and after project planning and implementation. It is still somehow remarkable that many large water sector investments are still undertaken without the necessary and proper institutional and financial conditions in place to ensure long-term financial, institutional, and social sustainability. There are numerous examples of water sector investments, where technical priorities override economic, institutional, and social concerns rather than the other way around. The same can properly be concluded for other types of investments.

In Vietnam there are still few examples of farmer- or community-based water user groups in charge of operation and management of irrigation or water supply systems. Many of the provincial government irrigation companies are still in charge of irrigation systems that ought to be fully handed over to farmers and users. The same applies to water supply, where the provincial water supply companies are often in charge of a series of water supplies scattered throughout the province. A proper decentralization of ownership and autonomy in operation and management still has to be developed in order to create effective water management systems in Vietnam. A major challenge for the Vietnamese government and its partners is therefore to develop and endorse simple, well-structured, and operational legislation that will enable local governments to work for a general and consistent move away from state management towards farmer and community-managed water systems. A key issue is to get water user groups legalized so that they become fully recognized entities with entrusted roles and responsibilities for planning, implementation, operation, and management, and with abilities to access finance. For example, pilot activities have been undertaken for developing and testing models for participatory irrigation management in selected districts in Dak Lak. However, more needs to be done for legalizing water user groups or associations, to develop their full role and responsibility for operation and maintenance, and their right and ability to collect and use water fees for such purposes (Danida, 2001).

Effective Water Governance

It is widely recognized that without appropriate policies, institutions cannot function, without appropriate institutions, policies will not work, and without a working set of policies and institutions, management tools are irrelevant. Without good governance civil society will not support the policies and will have a difficult time achieving sustainable and equitable water use. Good governance requires, above all, transparency and accountability by the institutions and participation by the citizens (Global Water Partnership, 2002). Good and effective governance is at the heart of IWRM.

The government approved a master programme in September 2001 for public administration reforms (PAR), covering four related areas:

organizational restructuring, human resources reforms, institutional improvement, and public financial management. Major challenges for the government include redefinition of roles, functions, and organizational structures of public institutions, including significant downsizing of many public institutions as well as human resources development and management to improve the quality of public servants. Salary reforms and improvement of general employment conditions are also believed to be among the key challenges for the government in the years to come. There is also scope and significant challenges lying ahead of the government for reforms of financial management mechanisms in the public sector, and improvements of the process of law making and of developing and issuing legal documents (World Bank, 2002a).

Most of the challenges facing the government and the development partners in the water resources and environment sectors in Vietnam are centred around improvement of governance, public administrative reforms, organizational restructuring, and human resources development. The government's fight against corruption in the administration and the society might also in the long-term contribute to the development towards a more effective water governance system in the country.

Also, among the main challenges of working towards better water governance in Vietnam are the needs to improve coordination and collaboration across sectors and traditional administrative boundaries; improved systems and practices for sharing of knowledge, information, and data; and to continuously support the development and movement towards an enabling sector framework and environment for integrated water resources and environmental management. Otherwise, many water and environment policies and strategies will remain the work of just good intentions.

Aid Management in Integrated Water Resources Management

During recent years, a number of donors have expressed increasing concern about high transaction costs, fragmentation of administrative capacity, and reduction in aid effectiveness caused by multiplicity of donor agencies' operational policies, procedures, and practices (World Bank, 2002b). Some of these symptoms are also identified in development assistance to water resources and environmental management in Vietnam. Various donors have a variety of arrangements for managing their aid programmes, also within water resources and environmental management. The present national system for managing ODA in Vietnam, as stipulated in Decree No. 17 from Ministry of Planning and Investment (MPI), is primarily based and organized around the so-called project approach, whereas more comprehensive and holistic sector-wide approaches still have to be developed and integrated into the existing Vietnamese administration (GoV, 2001). It is common practice that the government establishes semi-autonomous project management units, most often as temporary units and in parallel with existing institutions. There is a risk that the present model applied by the government for development assistance can lead to rivalries and duplication, and can undermine efforts to build capacity in permanent institutions (World Bank, 2002). To ensure that development assistance becomes an effective contributor for achieving the vision of integrated water resources and environmental management in Vietnam, it will be necessary to rethink and reorient the present project approach and hence work more effectively towards a more comprehensive, holistic, and integrated approach to development assistance in these sectors. A major challenge for the government and the central authorities in Vietnam will therefore be to cooperate with donors to rethink and reorient the present aid management system, which is primarily based on a traditional project approach rather than on sector-wide approaches. It will also be necessary to decentralize more of the development assistance to line ministries, provinces, districts, and communes.

As argued in this chapter, integrated water resources and environmental management is to a large extent concerned with addressing issues such as legislation, policy and strategy development, institutional and capacity development, human resources development and management, advocacy and dissemination, and other issues. Certainly there are also many technical issues to be addressed. However, the policy, institutional, and human resource issues are by far the most important issues. It is still a common understanding and view of many government institutions and officers that water resources and environmental management is equivalent with infrastructure development, delivery of equipment and so forth. There is a significant scope for developing the understanding, thinking, and practice of the basic and fundamental principles of water resources and environmental management. Many professionals in the water sector are often confronted with the traditional view and understanding that water resources management is more or less equal to infrastructure development or water service delivery. It is vital with support to legislation, policy and strategy development, institutional and capacity development, human resources development as well as advocacy and dissemination of best practices. In this context, the role and importance of national and international technical assistance from the private sector should not be underestimated in the water resources and environment sectors in Vietnam.

Working towards aid harmonization in water resources and environmental management might, in some cases improve the effectiveness of partnerships by eliminating unnecessary differences in operational requirements and reducing transaction costs of development assistance as well as help strengthen the national aid management system.

Conclusions

Many interesting and encouraging initiatives have been made in recent years in water resources and environmental management in Vietnam, and many national institutions and professionals profess good and impressive technical know-how and expertise as basis for their involvement and contributions in the development process towards integrated water resources and environmental management in the country.

Vietnam has a sound legal framework for water resources and environmental management. The LWR is an important part of the overall and general framework for water resources management in Vietnam. However, a major challenge for the government is to initiate the necessary reforms for putting the principles as outlined in the LWR into practice. It is a key challenge for the government and its administration to introduce, disseminate, and advocate for the enforcement of the LWR at local government levels and among water users.

If Vietnam is to develop a complete framework for effective water resources and environmental management, issues of good and effective governance will need to be addressed in parallel with public sector reforms. It will be necessary to undertake restructuring of a number of redundant government institutions dealing with water resources and environmental management. Public administrative reforms are necessary to downsize government institutions and to develop and maintain qualified staff through better salary structures and good working conditions. It will also be necessary to abolish a number of public water and environment institutions, and instead to encourage and actively support professional associations, private consultancies, contractors, suppliers and the private sector, in general, to establish themselves as new important partners in the water and environment sectors in Vietnam.

In the years ahead, it will be necessary to unlock the present public sector dominance in the water and environment sectors, and hence to allow for the inclusion of professional associations, non-governmental organizations, interest and water user groups, and other important development partners, not the least, the private sector. Along with developing the enabling environment for the sectors that will allow development partners to take up their intended roles and responsibilities and to perform accordingly, a reorientation of the state and public institutions and hence their role and function in the water and environment sectors in Vietnam is deemed necessary. The public institutions should be re-oriented to become policy makers, regulators, and monitors rather than implementers. An approach of 'no more business as usual' should be put on the agenda. Furthermore, there is a need to rethink and reorient the present aid management system, which primarily is based on a traditional project approach rather than on sector-wide approaches. It will be necessary to decentralize more of the decision making on development assistance to line ministries, provinces, districts, and communes. A key issue is also to get water user groups legalized so that they become fully recognized entities with entrusted roles and responsibilities for planning, implementation, operation, and management, and with abilities to access finance.

A move in the direction of introducing good governance and developing a renewed public sector must be supported by the development of new and modern higher education programmes in multi-disciplinary and cross-thematic water and environmental sciences. New and reoriented higher education programmes should focus on both new curricula development and modern teaching methodologies as basis for creating new and talented graduates to the public and private sectors.

In the long term, there is a need to develop independent water and environment institutions, which will be in a better position to provide objective and professional advice and guidance to the government, decision makers, civil servants, the public, and civil society. There is considerable scope for enhancing the involvement of private sector and civil society in the water resources and environment sectors as well as in the general development process across other sectors.

Integrated water resources management is in its infancy, and the future, successful achievements of integrated water resources management in Vietnam will very much depend on the general democratization and social development in the country as well as on the government's willingness and new initiatives to undertake effective reforms for good governance, public sector administrative reforms and restructuring, human resources development, and reorientation of the higher education system.

References

'Action Plan for Water Resources Development, Upper Srepok Basin Phase II', Vietnam National Mekong Committee, Dak Lak Provincial Peoples Committee, Mekong River Commission, Danida, April 1994.

'Action Plan for Water Resources Development, Upper Srepok Basin Phase III', Vietnam National Mekong Committee, Dak Lak Provincial Peoples Committee, Mekong River Commission, Danida, December 1999.

'Comprehensive Harmonization Program for the Government of Vietnam', Draft Issues Note and Action Plan, Operations and Country Services: World Bank, 2002b.

'Decree No 17/2001 on Regulation of the Management and Use of Official Development Aid', Hanoi: Government of Vietnam, 2001.

Hansen, J., 'Some Experiences with Implementing Urban Sanitation Projects in Vietnam', Technical Papers of International Seminar on Urban Sewerage in Vietnam, Hanoi: Vietnam Water Supply and Sewerage Association, December 2001.

'Institutional Assessment Study', Danida and MARD, 1999. February 1999.

'Integrated Water Resources Management', Technical Advisory Committee, TAC Background Paper No. 4: Global Water Partnership, 2000.

'Law on Water Resources', 20 May 1998, Decree, 179/1999/ND-CP of 10 July 1999 on implementation of the law on water resources, Government of Vietnam, 1998b.

'Main Findings of Project Completion Review Mission', Red River Delta Water Resources Sector Project: ADB 2002.

'National Law on Environmental Protection' 27 December, 1993, Decree No. 175-CP of 18 October 1994 detailing the implementation of the law on protection of the environment, Government of Vietnam, 1993.

'National Rural Water Supply and Sanitation Strategy', Vietnam: MARD and MoC, 1999.

'National Rural Water Supply and Sanitation Strategy', Vietnam: MARD and MoC, 2000.

'National Water Workshop: Proceedings', Hanoi: MARD, VUSTA, GWP, Royal Danish Embassy and Royal Netherlands Embassy, 2001.

'Preliminary Assessment of Water Resources and Water Resources Management Issues', Working Paper prepared for the National Water Resources Council, Vitenam: MARD, 2000(b).

'Project Completion Report for Support to Water Resources Management in Dak Lak', Support to Water Resources Management in Dak Lak: Danida, 2001.

'Regulation on Organisation and Operation of the National Water Resources Council', Attached with the Decision No. 67/2000/QD-TTg dated 15 June 2000 issued by the Prime Minister, Vitenam: MARD, 2000a.

'Regulations on the Exercise of Democracy in Communes', May 1998, issued in conjunction with Decree No. 29/1998/ND-CP, Government of Vietnam, 1998a.

Solanes, M., 'Report on the Law on Water Resources', Report prepared for Water SPS: Danida, 2001.

'The Comprehensive Poverty Reduction and Growth Strategy (CPRGS)', Approved by the Prime Minister as document no. 2685/VPCP-QHQT, 21 May 2002.

'Toolbox for Integrated Water Resources Management': Global Water Partnership, March 2002.

'Vietnam 2010: Entering the 21st Century': World Bank, 2000.

'Vietnam Development Report 2002', World Bank 2002(a).

'Vietnam Environment Monitor 2002', World Bank, Danida, National Environment Agency, 2002.

'Vietnam Water Resources Sector Review': World Bank et al., 1996.

'Water Sector Programme Support Document': Danida and MARD, April 2000.

9

INTEGRATED WATER RESOURCE MANAGEMENT FOR THE MEKONG RIVER BASIN

Ian C. Campbell

Introduction

The Mekong river represents a unique challenge for integrated water resource management (IWRM) in South-east Asia. The basin includes parts of six countries with a large proportion of the population poor and in many cases dependent entirely or in part directly on natural ecosystems for their livelihoods. While there has been a cooperative approach towards attracting development proposals among the lower Mekong basin countries since 1957, it is only since the signing of the 1995 agreement that Cambodia, Laos, Thailand, and Vietnam have broadened their cooperation to incorporate planning and sustainable management of the resource. The challenge in the Mekong, therefore, has two components: development of sufficient institutional integration and cooperation between countries for effective management, and developing the water resource to alleviate poverty within the basin while maintaining the natural ecosystems on which so many of the residents depend.

The Mekong Basin

The Mekong is one of the world's large rivers. In terms of discharge, Dudgeon (2000), citing Milliman and Meade (1983) and Van der Leeden (1975), ranked the Mekong fifteenth largest, but the value he cites for the mean discharge (11,048 m³/s) is considerably lower than more recent estimates. Mekong River Commission (MRC, 1997) gives a mean discharge of 15,060 m³/s, and ranks the Mekong eighth largest. The data for large rivers in developing regions is likely to be imperfect and in the short term it is not possible to be definitive, but the Mekong is clearly among the twenty largest rivers in the world in terms of discharge and probably one of the ten largest.

The Mekong rises in eastern Tibet in China at an altitude of almost 5000 m, and drains south-east to the South China sea (Pantulu, 1986; Hori, 2000) (Figure 9.1). By the time the river reaches Simao port in southern Yunnan Province in China, about one-third of its length, it is at an altitude of only 370m. As this would indicate, the headwaters of the river, and many of its major tributaries, are mountainous, but from Vientiane downstream the river mainly flows through country of relatively low relief. From Stung Treng downstream the river is fringed with extensive areas which are subject to inundation during average high-flow seasons.

Discharge in the river follows a seasonal pattern. The flood arises from snowmelt in the Himalayas in the headwaters, with wet season monsoonal rains increasingly contributing as the flow moves downstream. The river peaks in October at Phnom Penh in Cambodia (Figure 9.2), about a month later than the peak reaches Chiang Saen in Thailand, which is closer to the headwaters. Water quality in the mainstream is generally good (MRC 2003). Suspended sediment concentrations have been falling over the past fifteen years at all mainstream sampling sites above Pakse, as has conductivity in tributaries in north-east Thailand, although both of these effects are probably more related to increased numbers of control structures rather than improved catchment land-use practice. The most severe water quality problems are occurring in the Mekong delta where nitrogen and phosphorus concentrations are already high and have been increasing at almost all MRC sampling sites since 1985.

The countries which share the Mekong waters include some of the poorest in the world. In actual dollar values mean per capita income for Cambodia in 1999 was only US$ 1361, for Laos US$ 1471 and Vietnam US$ 1861 after adjustment for differences in purchasing power of the national currencies in each case (UNDP, 2001). In 1999, these three countries were ranked at number 121, 131, and 101 respectively on the UNDP Human Development Index where rank 163 was the least developed country. Although Thailand is wealthy compared with others in the region, much of the Thai territory bordering the river is amongst the poorest and least developed in Thailand. In the Lower Mekong basin (LMB), which refers here to that part of the Mekong river catchment which lies within the borders of Laos, Thailand, Cambodia, and Vietnam, there is relatively little industrial development. The population of over sixty million people (MRC, 1997) is supported largely by subsistence agriculture based primarily on rice and wild-caught fish. Social surveys have determined that people in this region consume between 28 and 67 kg of fish per capita annually (Jensen, 2000). Based on this data the capture of fish from the river and its associated ecosystems amounts to some 1.5 to 2 million tonnes per year

Figure 9.1: Map of the Mekong river basin

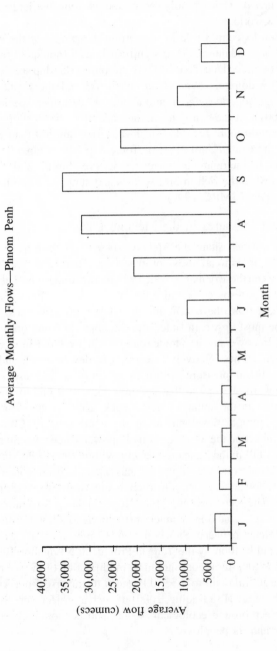

Figure 9.2: Average monthly discharge for the Mekong river at Phnom Penh

(Coates, in press). This possibly makes the Mekong the largest riverine fishery in the world.

Apart from its central role in supporting the people of the basin, the Mekong River is of international significance as a 'hot spot" for aquatic biodiversity (e.g., see Dudgeon, 2002). Well-known flagship species include the Irrawaddy dolphin (*Orcaella brevirostris*) and the Giant Mekong Catfish (*Pangasianodon gigas*), and about 700 described species of fish (Kottelat, 2001). At the family level, the Mekong is the most diverse river in the world with more recorded freshwater fish families than either the Amazon or the Congo/Zaire. Part of Tonle Sap lake is already listed by UNESCO as an international biosphere reserve (Smith, 2001), and the region is known to be rich in bird species and supports a number of rare mammalian species (MRC, 1997).

Development Pressures in the Mekong Basin

The countries which share the Mekong river basin have quite different long-term major national uses for the river. There are clear potential conflicts between these demands which will require trade-offs. The pressure on the river as a resource is still light, so there is still time to negotiate the trade-offs before the conflicting demands become serious.

China, the most upstream state, sees the upper Mekong primarily as a source of hydropower and as a trade route. South-western China, including Yunnan Province, has relatively low levels of development, and there is increasing political pressure within the country to increase efforts to develop the western areas so that they are not left behind by the rapidly developing east of the country (McCormack, 2001). Two major projects intended to promote development in the south-west of China would potentially impact on the Mekong. A hydropower scheme would make use of the 1900 m fall in the Lancang-Mekong within Yunnan Province (Hori, 2000). One reservoir, the Manwan, has already been constructed; a second, the Xiaowan, is under construction, and twelve more are proposed (Hori, 2000). The second scheme is the Upper Mekong Navigation Project which consists of removing channel obstructions to allow transit of ships up to 100 tonnes between Simao port in China and ports in northern Thailand. A publication by the Yunnan Provincial Navigation Bureau has argued that the long-term goal should be modification of the river channel to permit barge trains up to 4 × 500 tonnes to traverse from Yunnan to the South China sea (Liu Daqing, 2000). If such a project were to proceed it would have profound ecological, social, and economic consequences for the river and its people.

Myanmar, the next downstream state, controls a relatively small part of the basin within its territorial boundaries, and the Mekong basin makes up a relatively small proportion of the area of the country. Myanmar is the country with the least impact on, or interest in the Mekong. Possibly, use of water for irrigation would be the major likely use, but this is unlikely to be a significant impact in the context of the basin-wide usage patterns.

Laos also sees the Mekong primarily as a source of hydropower. Laos is landlocked with a low population density, making it difficult to develop export industries in areas such as manufacturing. More than 90 per cent of electricity in Laos is produced from hydroelectric plants (McCully, 2001) and hydro-generated electricity is seen as an export product which the government of the Lao PDR believes to have major growth potential with markets, initially, primarily in Thailand and Vietnam.

Thailand is primarily interested in the Mekong as a water source. A basin, by, basin review of Thailand's water resources concluded that while overall there was sufficient water, in several regions there was insufficient to meet demands (Kerdubon and Sethaputra, 1995). Two such regions are the central region around Bangkok and the north-eastern regions along the Mekong. These shortages have led to proposals such as the Kok-Ing-Nan proposal put forward by the Royal Irrigation Department which would divert water from three Mekong tributaries into the Chao Praya system which flows south to the central basin and Bangkok (Bangkok Post, 1998; Hori, 2000). There is also likely to be increased pressure to use the Mekong and its tributaries for irrigation in north-east Thailand, with one proposal, the Kong-Chi-Mun plan calling for the pumping of water via a 200 km aqueduct from the mainstream near Nong Khai to the Mun and Chi rivers (Hori, 2000).

For Cambodia, the main value of the Mekong is the fishery. For the Lower Mekong basin (including Thailand, Laos, Cambodia, and Vietnam) recent estimates put the annual wild-caught fish catch at 1.2 million tonnes per annum, of which almost 400,000 tonnes, valued at about US$ 400 million per year, is harvested in Cambodia (Jensen, 2000). In contrast to the wild-caught fish, aquaculture contributes only 200,000 tonnes of the total fish consumed in the basin as a whole and is probably an even smaller proportion of the fish consumed in Cambodia since most aquaculture activities are carried out in Vietnam and Thailand. More recent unpublished data suggest that the published estimates are still too low, but the relative proportions between wild-caught and aquaculture fisheries remain similar. Fish provides most of the protein and a large proportion of the calcium for the people of Cambodia (Jensen, 2001), so it is a key to the national welfare of the country. The other staple is rice, which is also dependent on the river as the source of water for both irrigated and naturally watered crops. Cambodia also uses the river for trade, with Phnom Penh serving as a port for ocean-going ships, but the sea trade is currently relatively small.

Vietnam relies on the Mekong for the water to support the rice crop in the Mekong delta—the source of 40 per cent of Vietnamese agricultural production (Osborne, 2000). Any alteration of upstream flow regimes is of concern to Vietnam, but of particular concern is any reduction in low flow volumes because seawater encroachment into the delta waterways is already impacting rice and vegetable (Joy et al., 1999) crop production. Vietnam is also utilizing Mekong tributaries as a source of hydroelectric power, with a new storage recently opened at Yali falls on the Se San River (CRES, 2001).

There is a clear conflict between some of these goals. Vietnam requires river discharge to be maintained to provide the water for rice, and Cambodia requires the water, and the annual flood to support the fish. On the other hand, Thailand would like to remove water from the river, and the Lao and Chinese hydropower developments will alter the flow regime and also reduce overall discharge.

Integrated Water Resource Management

Jonch-Clausen and Fugl (2001) identify three basic goals implicit in the definition of IWRM as given by GWP 2000: economic, social, and environmental. Economic efficiency of water use is necessary because of the increasing scarcity of water relative to demands. Social equity is a necessary recognition that all people have a right to access to adequate amounts of water of appropriate quality to sustain their well-being. Environmental and ecological sustainability is necessary to ensure that potential resource use by future generations is not compromised.

Integrated water resource management may be conducted at a range of scales. The scale of the river basin or aquifer is the most fundamental unit. However, there will frequently be cases where management at larger scales may be advantageous or necessary. For example, in cases where national boundaries encompass several river basins there may be decisions that not every part of the country will be self-sufficient in food. This implies that some regions would import food and the 'virtual water' required for its production (Jonch-Clausen and Fugl, 2001). This pattern of development is used for many large cities located in small river basins, for example. Therefore, integrated water resources management incorporates but is not synonymous with integrated catchment management.

The integration required for integrated water resources management is twofold. On the one hand there is a requirement to integrate information about the natural system—the water resources, agricultural, and ecological systems. Equally, there is a need for an integrated institutional response since many of the natural systems to be managed are the responsibilities of different national agencies. Most countries have distinct ministries with responsibilities for water resources, agriculture, and environment to name just three critical components of the natural system. In international river basins there is also an obvious need to ensure integration between national governments. Unless upstream and downstream countries develop compatible management and development strategies there is a likelihood of inefficient resource use or damage to the resource or even international conflict.

With the clear need for development within the Mekong basin to help alleviate the poverty, potentially conflicting resource development goals among the riparian countries, and a heavy dependence on the natural ecology by the poorest in the basin, integrated water resource management is essential for the Mekong. Clearly, integrated management of natural systems and integration of institutional systems both present great challenges within the basin.

Managing the Natural System

Natural systems that are not understood cannot be managed effectively, but understanding of natural systems is still not very well developed. In general, human understanding bears an inverse relationship to the complexity of the system. Thus, we understand how agricultural systems work better than we understand natural ecosystems.

In integrated water resources management one of the current weaknesses is a poor understanding of the functioning of river and stream ecosystems and their dependence on flow regimes. Clearly, the organisms living in streams respond to the aquatic environment at a number of different temporal and spatial scales. Precisely what environmental cues they respond to has been the subject of great debate in the ecological literature (e.g. Statzner and Higler, 1986; Davis and Barmuta, 1989) but it seems likely that different taxonomic groups are responding to different cues, and that a single taxon may respond to cues at several different scales.

Some researchers have argued that natural changes in the flow of a river, spates, floods and droughts, may be thought of as disturbances to the biotic community of the river (e.g. Lake et al., 1989; Downes et al., 1998). These researchers view the riverine community as essentially stochastic, dependent on such disturbance to maintain the biological diversity. That view is not universally accepted, but whether or not flow variation is viewed as a disturbance it is clearly necessary to maintain a healthy river. Reductions in flow variability have often been linked to invasions by exotic species of fish, for example.

Early research on the consequences of regulating rivers focused specifically on the consequences of reduced amounts of water and paid little attention to the change to the flow variability. The outcome of this research were recommendations for 'minimum flows' usually identified as the amount of water required to maintain fish such as trout within a river (e.g., Tennant, 1976). The minimum flow approach proved spectacularly ineffective even at maintaining the target species and was soon modified. One problem that became apparent early was that in streams with constant flows fine silt accumulated in the gravel beds required by trout for spawning. The silt reduced water flow through the gravel, and as a result oxygen concentrations also fell, and the eggs were suffocated. Within a few years, streams in which the 'minimum flow' had been maintained were almost devoid of fish.

Since that time a better appreciation of the influence of flow variability has begun to influence the environmental flow discussion. King et al. (2000) for example have identified a number of characteristics of natural flow regimes that seem to be critically important to maintain natural systems in rivers, and to incorporate these characteristics into environmental flow scenarios. For example, the vertical distribution patterns of much of the indigenous riparian vegetation along rivers in Zimbabwe was directly correlated with the water levels associated with particular flow events. Many aquatic plants require inundation all through the year but water shallow enough so that their flowers can emerge into air to permit pollination. As a result they cannot grow in deeper parts of the channel, and their distribution will be sensitive to increases in dry season low flows, because those members of the population unable to flower in air will be unable to produce seed. Other species grow in areas that flood less than once per two years but at least once in twenty or thirty years. Such species are sensitive to changes in wet season flows—if they are flooded too often, seedling survival is reduced but if they are flooded too rarely either seedlings fail to germinate or survival of mature individuals is reduced.

Unfortunately, little of this understanding has yet been absorbed by water resource engineers or within the IWRM literature. Too often, reduction of wet season flows and increasing dry season flows on rivers are seen as a good thing because they may reduce flooding and provide more dry season water for irrigation use and navigation. Chinese officials have been quoted in the press as intending to maintain a constant year-round flow in the upper Mekong once the Xiaowan dam is completed. If this were to occur it would have severe impacts on the riverine ecology for an appreciable distance downstream.

Courses in IWRM within the Mekong region seem to be exclusively taught by engineers and rarely if ever extend far beyond the application of large-scale hydrological models to river basins. Modelling has an important, but overrated, role in integrated management. Modelling allows the prediction of the hydrological consequences of river basin management activities, which is important. But what is more important is to be able to answer the 'so what' question that follows. What are the ecological consequences of those hydrological changes and, more importantly, what are the livelihood consequences? These are the key questions which cannot be answered using modelling and which tend to be ignored in IWRM courses.

The lack of ecological understanding of rivers is particularly acute in the countries of the Mekong basin. The major research effort has been focused on the freshwater fish which are so central to livelihoods in the region. Significant progress has been made in taxonomy (e.g., see Rainboth, 1996 and Kottelet, 2001) and in mapping migration patterns (e.g., see Poulsen and Valbo-Jorgensen, 2000) but relatively little work on food relations and none on determining migration cues. For other components of the biota the situation is far worse. There has been some work on human parasites and their hosts, but mainly from an epidemiological rather than an ecological perspective. Other than that the invertebrates have been very poorly served.

Campbell and Parnrong (2000) noted that very little of the published literature on freshwater science in Thailand concerned ecology; the emphasis was on taxonomy and species lists for particular localities. Research on riverine fisheries is being conducted in Laos and Cambodia, but there is no active research in either country on any other aspect of river ecology. Even in China there is very little published research on river ecology, with the notable exception of the work of Dudgeon and co-workers in Hong Kong (e.g., Dudgeon, 2000).

The paucity of river ecologists within the Mekong river basin has important consequences beyond the lack of understanding of the ecology of the system. Those designing IWRM courses for this region do not have ready access to sources of local information or expertise on the impacts of water resource development on the natural environment, so it is difficult for students to receive training in this area. Similarly, those who propose or evaluate developments have poor access to expertise in river ecology to assist them.

Integrating Institutional Management

Integrating institutional management also presents a challenge within the Mekong basin. Within the riparian countries there is a challenge to integrate management between government agencies both 'vertically' between national and provincial and local government levels and 'horizontally' between ministries and sectors. In countries such as Cambodia many of the institutions are still young and relatively undeveloped. In these circumstances, a period of institutional strengthening and development is to be expected. In all the countries of the basin, concern about the environmental consequences of development, and planning and management to minimize and ameliorate undesirable impacts have received emphasis only of late. However, it will take some time for the institutional responsibilities and boundaries to be completely delineated in practice.

The establishment of the Mekong River Commission following the international agreement signed in 1995 is a major milestone in cooperative international river basin management. The agreement established goals for river basin management for a river prior to problems or conflicts arising, or the river becoming irreparably degraded. That the agreement was reached in a region which has until recently been wracked by long and bitter war with the participating countries recent adversaries makes it even more remarkable.

The 1995 agreement provides a framework for future cooperation between the four countries that make the Lower Mekong Basin. It builds on the experience shared by the countries through the operation of the Mekong Committee, formed under the auspices of ESCAP in 1957, and the subsequent interim Mekong Committee. However, while its predecessors' roles were essentially to attract development projects to the member countries, the current MRC has a very different role.

The major roles for the MRC identified in the 1995 agreement are in basin-wide planning (Article 2), environmental protection (Articles 3 and 7), facilitation of equitable water use (Articles 5 and 6), and navigation (Article 9). This is a very different role than that of the Mekong Committee which was closely identified in the minds of many with proposed large-scale hydroelectric and irrigation schemes. Confusion about the difference between the role of the MRC and its forerunners still leads to some critics castigating the commission for the policies of the Mekong Committee with which they disagree. Others criticize MRC for not continuing with the role of the Mekong Committees.

Otherwise, ultimately, the success of the MRC as river basin manager depends on the cooperation of the four member countries, and the extent to which they can engage the upper basin countries, most notably China, in their vision. The one weakness in the present management structure for the basin is the absence of China as a participant. Myanmar is a far less significant player, because the proportion of the basin within Myanmar territory is small (3 per cent) and the basin in turn represents a small proportion of Myanmar's territory. China controls 21 per cent of the basin and produces 16 per cent of the discharge. While the impact China can have on the river as a whole is limited, its potential impact on the upstream Lao and Thai sections of the river is significant.

In recent times there has been increased engagement by China with the countries of the Lower Mekong basin. Both China and Myanmar participate as dialogue members at meetings of the joint committee of the Mekong River Commission. This participation has been increasingly active over time. Furthermore, the Chinese government has been actively engaging the LMB countries in discussions on the Lancang-Upper Mekong Navigation Project. However, there has not yet been significant dialogue on the potential ecological consequences of the proposed and current Chinese hydroelectric development on the upper Mekong river, even though recently completed projects such as Manwan dam are already having impacts on the lower river.

The Future of Basin Planning in the Mekong

The MRC has embarked upon a process of basin planning for the Mekong which incorporates outputs from both MRC and national programmes. A Basin Development Planning (BDP) group is working with ten sub-areas within the basin to identify assets and development options. The identification process is explicitly participatory involving input from local, provincial, and national-level groups and agencies. The process will include screening of national development proposals to identify those which have trans-boundary significance and are consistent with the sustainable development objectives of the four member countries.

The BDP is supported by a water utilization project (WUP) which has several components. The first component includes a hydrological modelling package to allow prediction of the hydrological consequences of proposed development options. The second component includes an interim basin flow management component which is attempting to identify the subsistence and ecological consequences of hydrological changes. A third component is developing basin-wide rules on data sharing, water use, and water quality to facilitate environmental management.

The third core programme at the MRC is the environment programme (EP). Among other things, the EP is collaborating with the WUP to conduct the environmental flow studies and also operates water quality and ecological health monitoring programmes within the basin. This is the programme which will provide the feedback on the success or otherwise of environmental management within the river system.

In order to manage water resources within the basin it is envisaged that the three core programmes would interact with each other, other stakeholders, and decision makers (Figure 9.3). While the present BDP is a limited-life project, it is intended to develop a continuing basin planning process. Similarly, the hydrological models and knowledge-base developed under the WUP project would continue to be maintained and utilized both at the MRC secretariat and national government line agencies. The decision makers who would use the outputs of the process would potentially include government planning agencies and sector agencies as well as international funding and aid agencies such as the World Bank, the Asian Development Bank, and national aid agencies.

Conclusions

Many of the people of the lower Mekong countries are extremely poor, and their economic and social situation will not improve without development in the basin. However, the poorest people in the basin are also those most dependent on the wild resources of the river, fish and aquatic plants, so if they are to benefit, development must ensure the maintenance of those resources. The Mekong basin must be managed as an integrated unit to ensure that development is sustainable and equitable, but integrated management will require much greater, and more sophisticated consideration of ecological issues than has traditionally been the case in integrated watershed management.

Many past development proposals in the Mekong and elsewhere have been promoted by strong sectoral interests. This has often resulted in insufficient attention being paid to the possible impacts on other interests or groups in the community. It is essential that planning procedures be developed which are far more inclusive so that the full range of sectoral and community interests are considered. The Basin Development Plan developed under the MRC is attempting to develop such a process.

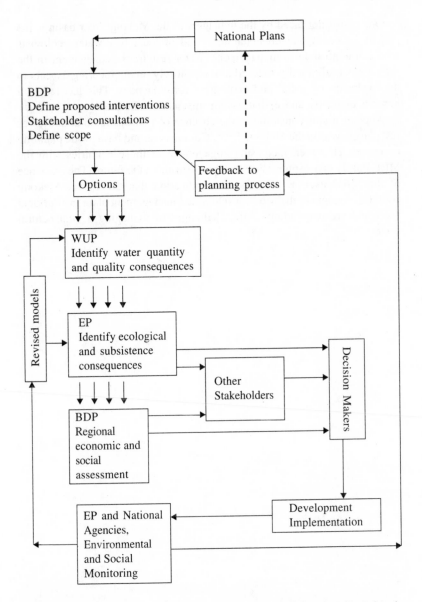

Figure 9.3: Flow diagram indicating information pathways proposed for the incorporation of environmental flows information into the MRC basin development planning process

One constraint faced by the managers of the Mekong river basin is the paucity of ecological understanding, and of local freshwater ecologists who can contribute to management. Water resources management in the region has traditionally focused almost entirely on engineering aspects to the exclusion of social and ecological consequences. This has led to a number of costly and embarrassing mistakes in the past.

Another major constraint is the absence of two upstream countries as full players within the Mekong River Commission and basin-wide planning processes. However, increasing engagement of those countries with the MRC family gives some cause for optimism about the future. The existence of the MRC itself is an important indication that the Lower Mekong countries recognize the need for integrated management at an institutional level and are responding to the challenges of managing a vital natural system.

References

Bangkok Post, 'Many rivers to cross', *Bangkok Post*, 30 December 1996.

Campbell, I.C., and S. Parnrong, 'Limnology in Thailand: Present status and future needs', *Verein. Internat. Verein Limnol*, 27, 2000, pp. 1–7.

Coates, D., (in press) 'Biodiversity and Fisheries management opportunities in the Mekong River Basin', Technical Development Paper, Mekong River Commission.

Davis, J.A. and L.A. Barmuta, 'An ecologically useful classification of mean and nearbed flows in streams and rivers', *Freshwater Biology*, 21: 1989, pp. 271–282.

Downes, B.J., P.S. Lake, A. Glaister, and J.A. Webb, 'Scales and frequencies of disturbances: rock size, bed packing and variation among upland streams', *Freshwater Biology*, 40, 1998, pp. 625–639.

Dudgeon, D., 'The most endangered ecosystems in the world?—Conservation of riverine biodiversity in Asia', *Ver. Internat. Verein Limnol*, 28: 2002, pp. 59–68.

Dudgeon, D., *Tropical Asian Streams: Zoobenthos, Ecology and Conservation*: Hong Kong University Press, 2000.

Hori, H., *The Mekong: Environment and Development*, Tokyo: United Nations University Press, pp. 398.

'Human Development Report', New York: UNDP, 2000.

'Integrated Water Resources Management', TAC Background Paper No. 4, Stockholm: GWPTAC (2000).

Jensen, J., '1,000,000 tonnes of fish from the Mekong?' *Mekong Fish Catch and Culture*, 2(1), 1996, pp. 1–12.

Jensen, J., 'Can this really be true? Rice yes and fish please!' *Mekong Fish Catch and Culture*, 5 (3), 2000, pp. 1–3.

Jensen, J.G., 'Traditional fish products: the milk of Southeast Asia', *Mekong Fish Catch and Culture*, 6 (4), 2001, pp. 1–16.

Jonch-Clausen, T. and J. Fugl, 'Firming up the conceptual basis of integrated water resources management', *International Journal of Water Resources Management*, 17, 2001, pp. 501–510.

Joy, C., G. Radesovich, T. Thuc, T. and P.X. Phuong, 'Water Management Case Study for the Mekong Delta in Vietnam', UNDP Project VIE/97/010 Hanoi: Vietnam National Mekong Committee, 1999.

Kerdubon, M., and S. Sethaputra, 'Status of water situation in Thailand's river basin', *Suranaree J. Sci. Technol*, 2, 1995, pp. 161–170.

King J.A., R.E. Tharme, and M. de Villiers (eds), 'Environmental Flows for Rivers: Manual for the Building Block Methodology', Pretoria: Water Research Commission Report 576/1/98, 2000, pp. 452.

Kottelat, M., *Fishes of Laos* Colombo: WHT Publications, pp. 198.

Lake, P.S., T.J. Doeg, R. Marchant, 'Effects of multiple disturbance on macroinvertebrate communities in the Acheron River, Victoria', *Australian Journal of Ecology*, 14, 1989, pp. 507–14.

Liu Daqing, 'Provisional plan for development of navigation and regulation of navigation channel on the Lancang-Mekong River', Paper from Sub-Regional Workshop on Technological Development of Inland Water Transport Infrastructure, Kunming, China: ESCAP, 2000.

'Mekong River Basin Diagnostic Study, Final report', Bangkok, Thailand: Mekong River Commission, 1997.

McCormack, G., 'Water margins: Competing paradigms in China', *Critical Asian Studies*, 33, 2001, pp. 5–30.

McCully, P., *Silenced Rivers: The Ecology and Politics of Large Dams*, London and New York: Zed Books, 2001, pp. 359.

Milliman, J.D., and R.H. Meade, 'World-wide delivery of river sediment to the oceans', *J. Geol*, 91, 1983, pp. 1–21.

Osborne, M., *The Mekong: Turbulent Past, Uncertain Future* St Leonards, Australia: Allen and Unwin, 2000, pp. 295.

Pantulu, V.R. 'The Mekong River System', in Davies, B.R. and K.F. Walker (eds). *The Ecology of River Systems*, Monographiae Biologicea 60, Dordrecht, The Netherland: Dr W. Junk Publishers, 1986, pp. 695–719.

Poulsen, A.F., and J. Valbo-Jorgensen (eds), 'Fish migrations and spawning habits in the Mekong mainstream: a survey using local knowledge', AMFC Technical Report, Phnom Penh: Mekong River Commission, 2000.

Rainboth, W.J., *Fishes of the Cambodian Mekong*, Rome: Food and Agriculture Organization, 1996.

'Study into the impact of Yali Falls Dam on resettled and downstream communities', Hanoi: Vietnam National University: Center for Natural Resources and Environmental Studies (CRES), 2001.

Smith, J., 'Biodiversity: The life of Cambodia, Cambodia's Biodiversity Status Report—2001', Cambodia Biodiversity Enabling Activity, Phnom Penh, Cambodia, 2001.

Tennant, D.L., 'Instream flow regimens for fish, wildlife, recreation and related environmental resources', *Fisheries*', 1, 1976, pp. 6–10.

Van der Leeden, F., *Water Resources of the World*, New York: Water Information Centre. Inc., 2001.

10

INTEGRATED WATER RESOURCES MANAGEMENT IN CHINA[1]

Katri Makkonen[2]

Introduction

The purpose of the analysis of China's role in integrated water resources management in South and South-east Asia is to analyse how integrated water resources management (IWRM) is practised in China, particularly in terms of international watersheds. The aim is to obtain a comprehensive picture of the reasons behind China's current practices to cooperate and suggest future approaches, which could facilitate the integrated management of the trans-boundary rivers. Whether China cooperates more in the future on the regional management of water resources remains a question that will greatly affect the development of the entire area.

[1] A more detailed version of the analysis of China's role in integrated water resources management in South and South-east Asia can be seen in the Internet: http://www.water.hut.fi/wr/research/glob/pubications/makkonen/alku_v.html

[2] The study was partially funded by the Academy of Finland, within the project 45809, and partially by the Ministry of Foreign Affairs of Finland. I would like to express my gratitude and special thanks to Prof. Pertti Vakkilainen, Dr Olli Varis, Prof. Asit K. Biswas, Jukka Mehtonen, Timo Kuronen, Maria Suokko, Josef Margraf, Liminguo Margraf, Yang Zheng Bin, Chen Yongsong, Tommi Kajander, Jörgen Eriksson, Prof. He Daming, Feng Yan, Jock Whittlesey, Ass Prof. Haoming Huang, Dr Dajun Shen, Hein Mallee, Dr Eva Sternfeld, Hannu Toivola, Terttu-Liisa Aho, Prof. Xiangrong Wang, Prof. V.K. Damodaran, Prof. Tong Jiangdong, Prof. Zhongyuan Chen, Prof. Xu Jian Chu, Wang Wenzhang, Huang De Chang, Xu Shi Yin, Qu Wen Long, Cao Meng Liang, Dong Xue Jun, Li Ging, Li Fengling, Dr. Lailai Li, Prof. Guanghe Li, Juho Rissanen, Antti Rautavaara, Prof. Ji Zou, Irene Wettenhall, Osmo Tammela, Chantawong Montree, Juha Uitto, Jim Nickum, Sulan Chen, Ian Campbell, Dave Hubbel, Dorit Lehrack, and Dale Campbell.

What makes China's management and development of water resources so significant internationally? The country is huge, and the impacts of actions related, for example, to industrial development, transportation, or pollution cannot be framed inside one country. Furthermore, the fact that many trans-boundary rivers either originate from, or flow to, China makes water questions internationally important. This topic becomes even more sensitive when one notes that, in many of these international river basins, the availability of water is critical, and people are strongly dependent on the rivers. Many of the great rivers in South and South-east Asia, including Mekong, Salween, Irrawaddy, Brahmaputra, Red and Pearl, originate from the Tibet Autonomous Region (TAR) and China's southern provinces.

This chapter provides an overview of China's role in international water issues and of the factors affecting this role now and in the next decades. Important issues to consider are as follows:

   (i) Little information has been available on China's management of trans-boundary rivers.

  (ii) The country is a significant player in water resources management in South and South-east Asia because several great international rivers originate from China.

 (iii) Chinese society has been opening gradually to the outside world—open discussion both inside the country and with the neighbouring countries has hitherto been difficult. As a result, China has often stayed, or been left outside, discussions about regional cooperation in the water sector.

 (iv) China's willingness to cooperate in the management and development of water resources has been minimal so far, and the real reasons behind this are still not definitively known.

  (v) The plans China has, and is expected to have, for trans-boundary rivers have significant impacts on the livelihoods of the neighbouring countries in various, and often very profound ways.

By better understanding the complexity that lies behind the decisions made on international water sector, scenarios for the impacts of different future actions can be drawn. Based on the analysis, propositions are made to enhance international cooperation, aiming to improve economically, socially and environmentally sustainable development in the area.

River Basins of Interest

The rivers of interest for this analysis were selected for their geographical situation, international character, and great importance to the riparian countries. All the five great Himalayan rivers dealt with in this chapter originate from South or South-west China—either from the Qinghai-Tibet Plateau or the Yunnan province—and then flow towards the countries in South and South-east Asia. There are a few other international rivers originating in South and South-west China that are not dealt here, as their international character is not significant for China. These rivers include the Indus, which have only the very beginning of them in the Chinese side, as well as the Pearl River, which is mainly an internal Chinese river, having only certain tributaries in Vietnam.

LANCANG-MEKONG

The Mekong, Lancang as it is called while in China, runs through China, Laos, Thailand, Myanmar, Cambodia, and Vietnam. It is one of the world's largest rivers in terms of discharge, with an average of 15,000 m^3/s, and is the largest source of water in South-east Asia. Of the total length of 4880 km, 1161 km of the Lancang-Mekong is located within China. On its way, the river passes through very different climates, morphology, geology, culture, social and economic conditions and areas. The watercourse, soil coverage, and hydrological characteristics are all quite complex in the Lancang-Mekong.

The Lancang-Mekong river basin covers parts of six countries with a large population depending directly on natural ecosystems for their livelihoods. The catches of a huge number of species of fish are the major sources of protein throughout the basin. The river is important for rice production, particularly in the Mekong delta in Vietnam, and in the Tonle Sap lake in Cambodia, which is highly dependent on the annual floods of the river. For China, the Lancang is particularly important as a source of hydropower potential and for transportation.

SALWEEN

The Salween river, like the Lancang-Mekong, originates from the Qinghai-Tibet Plateau, and runs through China, Myanmar, and Thailand. It is relatively close to the Lancang-Mekong. The Salween flows several hundred kilometres through southern China before entering Myanmar, after which it forms the Myanmar-Thailand border for about 110 km. Flowing through eastern Myanmar, the river finally empties into the Andaman Sea.

The river is rich in water resources. However, owing to the dangerous rapids, the navigability of the river is restricted. For the time being, the major economic use of the river is floating teak logs from the forests of

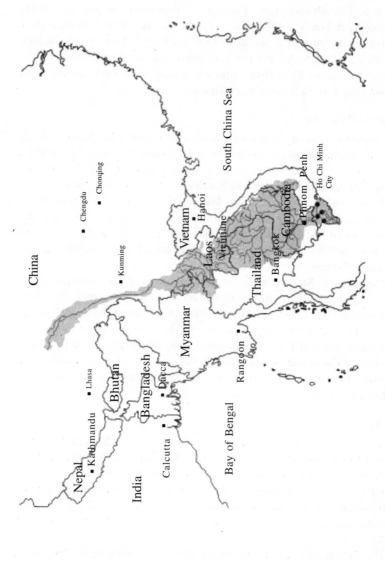

Figure 10.1: Lancang-Mekong river basin

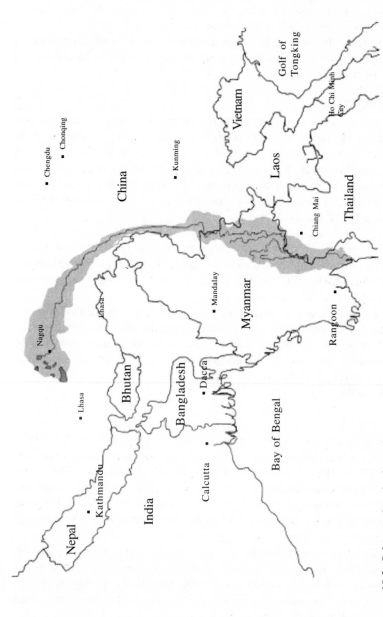

Figure 10.2: Salween river basin

south-eastern Myanmar to the sea. Although the catchment area of Salween is limited and sheltered from seasonal rains, its water volume fluctuates considerably from season to season.

Despite its huge hydropower potential, the Salween river remains untapped. Plans to develop the river are nevertheless in discussions in the riparian countries.

IRRAWADDY

With a total length of 2150 km, the upstream stretch of the Irrawaddy situated in China is only 178.6 km long. After passing the glaciers of the high and remote mountains of northern Myanmar, the river flows through western Myanmar. It drains the eastern slope of the country's western mountain chain, and then empties into the Andaman Sea, where it forms a considerable delta. The total area of Irrawaddy covers 431,000 km^2, of which 43,300 km^2 is in China.

One of the main characteristics of the river is the great variation of the discharge due to the monsoon rains between June and September. During the summer months, rapid melting of glaciers increases the flow. Mean low and flood discharge of 2300 m^3/s and 32,600 m^3/s respectively have been recorded near the head of the delta.

The water resources of Irrawaddy are particularly important to Myanmar, where economic development remains low, and dependence on agriculture and irrigation are high. With its base width of almost 200 km, the Irrawaddy delta is one of the world's major rice growing areas. For China, the importance of the river is not significant, as only a small stretch of it is in the country.

BRAHMAPUTRA

The Brahmaputra, Yarlung Zangbo in Chinese, originates from the northern slopes of the Himalayas in Tibet and then flows south out of China into the Bay of Bengal, passing India and Bangladesh on its way. The river is the biggest trans-Himalayan river system, having a total length of 2900 km. The total drainage area is 938,000 km^2, of which 239,200 km^2 is in China.

About 270 km south of the India-Bangladesh border, the Brahmaputra joins the Ganga River.

The river floods regularly, with both positive and negative impacts. For agriculture, the effect is positive as part of the sediment carried by the floods is stratified as a thin layer over the delta, thus renewing the fertility of the soils. The sedimentation also forms new islands on the delta, which are colonized by the farmers.

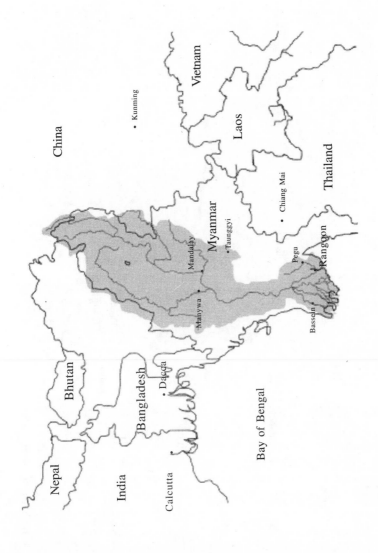

Figure 10.3: Irrawaddy river basin

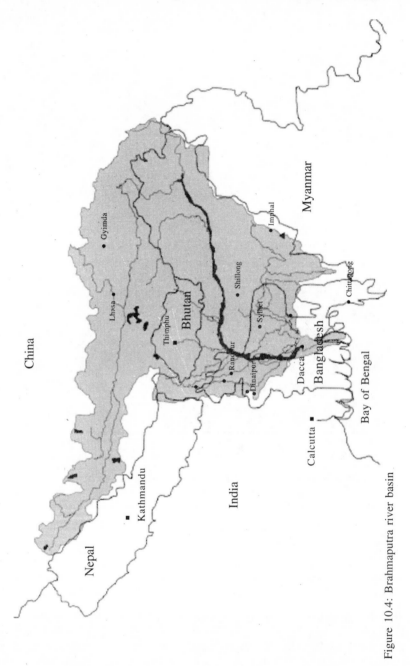

Figure 10.4: Brahmaputra river basin

The Brahmaputra is partly navigable at both ends, to Dibrugarh in India about 1100 km from the Bay of Bengal, and in Tibet for a distance of about 640 km. The need for river transport is particularly important to the continuing development of economic resources in the lower Brahmaputra valley.

The natural hydropower reserves in the Yarlung Zangbo stream and its five main tributaries are the second highest in China, after those of the Yangtze River, and are about 111,000 megawatts. The natural hydropower reserves per unit are the highest in China and three times that of the Yangtze River, being about 460 kW/km^2.

RED RIVER

The Red River, Yuan Chiang in Chinese, originates from the mountains of Yunnan Province in China and then flows through Vietnam to enter the South China Sea. Near the sea, it forms an extensive delta. The total length of the river is 1280 km, of which 667 km is in China. Of the total catchment area of 113,000 km^2, 74,000 km^2 is situated in China. About twenty-four million people live in the entire river basin, out of which seventeen million live in the delta in Vietnam.

The climate in the river basin varies from tropical to sub-tropical, and is dominated by the east-Asia monsoon, which causes significant seasonal variations in rainfall. The discharge of the Red river is highly variable, from a minimum recorded dry season discharge of 370 m^3/s, to a maximum of 38,000 m^3/s. The continuous floods in the river basin, caused by great amounts of river-borne silt, changed land use, and deforestation, pose a significant problem, especially in the delta region of Vietnam.

Key Questions

The principal questions for which answers have to be found are the following:
  (i) IWRM in China—is it a concept used in China, and if it is, what meaning does it have?
  (ii) What kind of plans does China have for the international rivers? What kinds of projects have been completed in recent years, and which are about to be completed in the near future?
  (iii) What has China's cooperation with its neighbouring countries been like in reality, both in China's and other countries' views?
  (iv) What are the driving factors behind China's willingness to cooperate? Assumptions made for the factors affecting the level of cooperation are as follows:

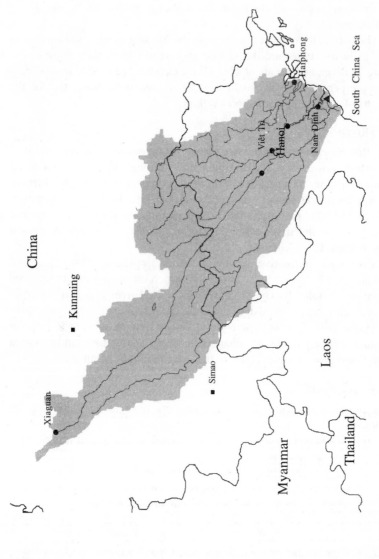

Figure 10.5: Red river basin

a) Structure of Chinese society and politics—central administration and China as a superpower;
b) Historical factors—relationships between the countries;
c) Strong economic development and pressure to develop further—the necessity of the projects targeted to water resources development;
d) Challenges inside the country and resources required to solve them, e.g., economics, funding, environment, governance, and capacity building; international issues may have a low priority;
e) Lack of adequate benefit—what China would really achieve through increased cooperation. This issue is bound to the geographical setting of most international basins that China shares since it is an upstream riparian;
(v) Is China willing to cooperate more in the future?
(vi) Which factors might increase China's willingness to cooperate or what could act as a motive for China? Attention is focused on the opening policy, increasing participation in the international community, and economic factors related to South-east Asia.

Findings

IWRM IN CHINA—IS IT A CONCEPT USED AND IF IT IS, WHAT MEANING DOES IT HAVE?

The answers are mainly based on interviews and discussions held in summer 2002 with Chinese water and environmental experts, government officials, NGO representatives, as well as international organizations' representatives working in China. The results of the interviews can be summarized in the following way:

(i) The term IWRM was recognized and it does appear in the government's official definitions and regulations concerning water management. However, it was admitted that the concept is still virtually unused in China. Integrated management only seems to exist in theory, and in most cases, the implementation is very poor or entirely lacking.
(ii) The idea of integrated management is to cover the entire river basin. This does not appear to be an attractive point of view to China, particularly in the case of trans-boundary rivers.
(iii) Practising integrated management in China includes two different aspects—integrating different forms of use and integrating the work of different governmental bodies and structures. With regard to the former, there is no serious lack of water in the river basins of

interest to this analysis (except some occasional shortages during the dry season), due to which there is no significant competition between agriculture and other forms of use. However, the need for water is growing in all the sectors, which will make the situation more difficult in the future. Industrial and economic development is given priority in most cases, before environmental aspects or sustainable development. When producing energy, it is significant for whom it is produced—for local use, to the special economical zones, or for export purposes. The needs of special economic zones and income from foreign trade seem to have a high priority.

When it comes to the second aspect of integration, cooperation between governmental structures and bodies is minimal. Institutions are competing with each other, which discourages cooperation. These problems take place at all levels: the ministries are not familiar with each other's work and overlaps often occur. The situation is the same at the provincial and local levels. Officials are often in a contradictory position—they are responsible to both local governance and to the central body of their specific field of study, which may have opposite goals.

WHAT KIND OF PLANS DOES CHINA HAVE FOR THE INTERNATIONAL RIVERS? WHAT KINDS OF PROJECTS HAVE BEEN COMPLETED IN RECENT YEARS, AND WHICH ARE ABOUT TO BE COMPLETED?

There is great pressure in China to develop the watersheds to support economic development. The main reasons for developing rivers flowing in south and south-west China are energy production and transportation. Developing these rivers is a part of China's strategy for transmitting electricity from resources-rich western areas to the power-short city of Shanghai, and equally power-short coastal and eastern provinces such as Guangdong and Jiangsu (Xinhua, 2002). Energy is also required to support local development. The export of energy is also important for foreign trade, and improving navigation plays a key role in areas where no comprehensive transportation infrastructure exists.

China has been accused of not negotiating with its neighbouring countries when planning and executing projects. China for its part argues that the impacts that the projects have over the borders are mainly positive, and thus no wider negotiation is required.

LANCANG-MEKONG—THE DAM CASCADE

China's plans to build a cascade of eight dams on Lancang are well-known. The plan is to take advantage of a 700 m 'drop' in the 750 km stretch in middle and lower Lancang (McCormack, 2000). Two large hydroelectric dams, Manwan and Dachaoshan, have been built in Yunnan Province, and construction of a third dam, the massive 290 m-high Xiaowan hydroelectric project, began in 2001.

The Manwan dam, completed in 1993, has the capacity to generate 1500 megawatts of electricity. The construction of the second dam, Dachaoshan, began in 1996, and the first section was brought into operation in 2001. The third one, the Xiaowan dam, will be China's second largest dam, smaller only to the Three Gorges project on the Yangtze River. The decision to build the dam was taken in 1999, and it is expected to take until 2012 to complete. The Xiaowan dam will have the effect of damming the river back for 169 km. The installed capacity of the dam will be 4200 megawatts, and the water storage capacity of 15 billion m^3. While giving priority to power generation, the dam will perform other functions such as flood control, irrigation, silt retention, and navigation. The next in line, the Jinghong dam, is at the feasibility study stage.

The Lancang dams are expected to supply power to China's industrial centres in the east, as well as to the growing markets in South-east Asia, particularly to Thailand. The dams are actually concentrated close to Myanmar and Laos, with relatively easy access to Vietnam and Thailand (McCormack, 2000). According to a memorandum signed between China and Thailand in 1998, these power stations will start providing electricity to Thailand from 2013. The Jinghong power plant plays a major role in the agreement, providing 80 per cent, or even more, of the energy it generates to Thailand (Wang, 2002; Cao Meng Liang et al., 2002).

The Lancang cascade will generate a maximum installed capacity in excess of 15,000 megawatts, annual energy output being about 70,000 gigawatt-hours (GWh). This means a power generation capacity equivalent to about 80 per cent of the Three Gorges dam (McCormack, 2000).

The question of impacts on lower reaches caused by China's dams remains unanswered as there have been many divergent opinions. China claims that studies here been carried out on the whole river basin (Feng, 2002). According to the research, the impacts would be mainly positive, as during the dry season the amount of water in the river could be increased and during the rainy season flood prevention will be improved. China's statements have been criticized heavily for underestimating the impacts. Even for the 292 m-high Xiaowan dam, the Chinese experts claim the impacts on the lower reaches are likely to be modest (Chen, 2002). With the Xiaowan power station, a reservoir with water storage of 15 billion m^3 will be built. This is expected to reduce the amount of water flowing downstream by 17 per cent during flood seasons and increase the flow by 40 per cent in dry seasons (He, 2002; Xinhua, 2002). Reduction of silt in the water would also help flood prevention. In addition, the Chinese argue that in all the cases, the effects would only reach Lao PDR, and not to the countries on the lower reaches.

Many regional and international specialists have different opinions, maintaining that the consequences will be severe and cause considerable human sufferings and environmental damages as the quality and quantity of the flow will change significantly. One fear is that, as the seasons reach different stretches of the long river at different times, the flow of water caused by the release of water from the dams during the monsoon season would actually create more flooding in the downstream. The changes in suspended sediment concentration in the downstream due to the upstream dams are likely to be significant. Trapping sediment may cause severe problems due to the unbalanced sediment concentration which may increase bank and bed erosion (Kummu et al., 2004). Additionally, the flux of nutritious sediments is extremely important for fisheries and agriculture production in the lower Mekong. Out of the total suspended sediment load of the Mekong basin, an estimated 50 per cent comes from China. After the completion of the Manwan dam in 1993, a significant reduction of the sediment flux in the downstream has already been reported (Kummu et al., 2004). Neighbouring countries claim that they have never seen an environmental impact assessment, and the existence of a document covering the entire river basin has been strongly questioned. China, however, claims that the problem is the language barrier as all their assessments and research are done in Chinese, and there are no funds for translations (He, 2002; Feng, 2002).

LANCANG-MEKONG—IMPROVING THE NAVIGATION

Another important issue for China is to improve navigation on the Lancang-Mekong, and consequently increase the trade between the neighbouring countries. The plan is to upgrade the channel of the Lancang-Mekong river to link China to the South-east Asian markets, following a regional commercial navigation agreement signed by China, Laos, Myanmar, and Thailand in 2000. The agreement involves blasting twenty-one natural rapids and shoals to clear the way for a channel to allow large cargo ships to navigate along the 886 km route between Simao Port in China and Luang Prabang in Laos round the year.

Cambodia and Vietnam claim that they were never asked or even properly informed about making the agreement, although they are the two countries most dependent on the Mekong River. The environmental impact assessment made for the project has been criticized for not assessing the project's potential impacts on the river's fisheries and food supply, or the economies of hundreds of fishing communities living along the Mekong (Yu Xiaogang, 2002).

It seems that the two richer countries of the area, Thailand and China, are the ones that will benefit most from the navigation improvement. For both, the lack of infrastructure required for transportation represents an essential focus for developing their national economies. However, even for Thailand, the benefits have been questioned.

It is thought that the project will change remarkably flow velocity thereby increasing soil erosion and floods, and disrupt the fish breeding grounds as a result of blasting away the rapids. On the lower reaches, millions of people are dependent on fishing as both a source of livelihood and nutrition. Once again, China considers the impacts not to be severe. It must be admitted though that the Lancang-Mekong, as it is a narrow and strongly turbulent river, is very difficult to navigate because of rapids, sharp curves, and shallow water in the dry season. At present, only cargo ships with 100 ton capacity are able to navigate the most difficult stretches of the river, which makes the effective use of the waterway difficult. The Lancang-Mekong will, after dredging, be able to handle 300 ton ships (Li Ging, 2002), with an annual cargo capacity that may reach as high as two million tons, and transport 400,000 people per year. As there are no proper alternatives for transportation, and as building roads and railways is extremely expensive in this area because of the soft soil and heavy rains that undermine the foundations, navigation has been claimed to be the only feasible way to transport goods. However, economic priorities, degradation of natural resources and the welfare of millions of people should all be considered concurrently.

SALWEEN AND IRRAWADDY

As a result of the remote location, the Salween and Irrawaddy rivers have not received as much international attention as the Mekong. For China, the two rivers are located far from the populated east coast centres. The Irrawaddy River has so far not been of specific interest to China, as the stretch of the river situated in China is short, less than 200 km. The great variations in discharges also make it difficult to generate hydropower. For Myanmar, the river is important as a means of transportation and source of fresh water. It is possible that two countries may jointly decide to develop the river in the future.

Until recently, it was thought that China's interest in the Salween was low. The river valley in the upper catchment area is long and narrow, which makes construction of dams difficult. High costs and technical difficulties might give the impression that significant development on the Chinese side of the Salween would be unlikely, and that the main interests in the development of the river would come from the two lower riparians (Paoletto & Uitto, 1996).

However, the Salween is rich in water resources, and might play a major role in energy generation. Serious plans to develop the river have recently appeared in all the riparian countries. China plans to build a cascade of thirteen dams in the Upper Salween, mainly in the Yunnan province (New York Times, 2004a). Myanmar and Thailand have agreed to build jointly three dams in the their sides of the river (BBC News Online, 2004).

China's plans have nevertheless been halted because of the strong public opposition within the country. The proposal became public in August 2004, when reports appeared in the Chinese media (New York Times, 2004b). The strongest supporters of the dams appear to be the local officials. Their motives are somewhat understandable and rather similar to the case of the Lancang-Mekong. The Yunnan province is among the country's poorest and needs energy both for its own development and for delivering income through power trade. The officials predict that the dams would help to provide jobs and raise incomes (New York Times, 2004a). The cascade, with an installed capacity of 20,000 megawatts, would also help to deal with the nationwide power shortage (BBC News Online, 2004).

The criticisms for developing the river are nevertheless strong. In addition, the Salween is one of the last pristine rivers in Asia. The area in China called Three Parallel Rivers where Yangtze, Mekong and Salween run parallel to one another, was designated as a UNESCO World Heritage Site in 2003. The dam opponents worry that the natural areas will be disrupted by the creation of a huge power project (BBC News Online, 2004). It has been estimated that the cascade would force the relocation of about 50,000 people (New York Times, 2004b). According to the opponents, the possible impacts in both China and downstream have not been adequately assessed (BBC News Online, 2004).

BRAHMAPUTRA

The Brahmaputra is situated in a remote area like many of the great Himalayan rivers. This environment makes it very challenging to develop the river, as a considerable sum of money would be required to complete the projects. The tensions between the two neighbours, China and India, due to historical events and competing positions as the great powers in South and South-east Asia, do not make it any easier to plan projects that may seriously affect the other nation.

For India and Bangladesh, the Ganga-Brahmaputra basin is a major issue. The number of people living in the area, the vulnerability of the flat-terrained Bangladesh to the floods, and water scarcity in the Ganga basin make the question critical.

Although there have been disputes between Bangladesh and India over India's earlier projects on the river, the Chinese side of the river has so far remained in a natural state. However, the river has a much higher capacity for hydropower potential than that of the mighty Three Gorges dam on the Yangtze River.

As the hydropower sources are rich, it may be possible to construct large projects on the Yarlung Zangbo River (Liu, 2001). The suggested location is around the Najabawa mountain peak, which is the highest peak in the lower section of the Himalayan mountains. The river makes almost a U-turn in this longest and deepest canyon in the world, with a length of 496.3 km and a depth of 5382 m. The canyon is partly very narrow, the narrowest point being only 75 m, making the water flow at a speed of about 10 m/s. The canyon area is also strongly precipitous. The 2600 m drop in the canyon is significant, providing a generating capacity of 68,800 mw. A dam having a normal water level of 2970 m, a head of 2340 m, and an average discharge of 1900 m^3/s is proposed near Pai. The installed capacity of this hydropower station would be 38,000 mw, which is twice the capacity of the Three Gorges hydroelectric power plant. The station would be the largest one of its kind in the world.

In some plans, the power plant would be executed by constructing a tunnel through the mountains near the Sino-Indian border, and letting the water flow through the tunnel. What makes it even more interesting is that in the area concerned, there have been border disputes between India and China in the past, and it is still not clear where the borderline really lies. This makes the proposal highly complicated.

RED RIVER

It has been speculated that the next major international river China may develop after the Lancang-Mekong could be the Red river in which China and Vietnam share a common interest. For China, the production of hydropower would mean more capacity for its industries, and Vietnam for its part would like to have more dams to control soil erosion. Less sedimentation would improve flood prevention in the area where serious floods occur regularly and threaten both human lives and property. Supporting navigation would also be beneficial for both the countries. The topic was discussed in 1998 when an international symposium was held between the two countries (He, 2002). Sino-Vietnamese relations seem to have improved remarkably in recent years, after a turbulent and hostile past due to conflicts and political disputes. China also shares common interests with its bordering country particularly in the trade sector, and it might not be surprising to see future cooperation between the two countries in developing the Red River.

WHAT HAS CHINA'S COOPERATION WITH ITS NEIGHBOURING COUNTRIES BEEN LIKE IN REALITY, BOTH FROM CHINA'S AND OTHER COUNTRIES' VIEWPOINTS?

The state of cooperation appears different depending on context and point of view. Different views on the topic exist inside China and among the neighbouring countries, as well as among international organizations involved in China. So far, China has signed only very few international agreements concerning the use of international water resources, but the situation seems to be changing. It is difficult to determine the real state of cooperation, but it is quite obvious that problems do exist with the current situation.

As regards China's cooperation across borders, different multilateral and regional levels can be distinguished. The directly neighbouring countries and the countries sharing the same river banks, the Mekong River Commission (MRC) and Greater Mekong Subregion (GMS) countries, form an important basis for the country's international relationships. The even larger nodes of cooperation include combinations of the ASEAN countries and their neighbours, the so-called '10 + 1' (ASEAN and China) and '10 + 3' (ASEAN and China, Japan, and South Korea).

The official Chinese view is that the level of cooperation is high, or at least that China seeks a mutual benefit when planning any projects. The image of China as a country seriously considering the international consequences of its actions, and willing to negotiate with other nations, is being drawn. As reasons for other countries' concern and experience of China's unwillingness for cooperation, government officials cite such factors as the language barrier and lack of resources for information exchange and common activities (He, 2002; Feng, 2002).

When we consider China's bordering countries, the relationship is complex because of governmental and societal views. The differences in economic development between South-east Asian countries are particularly important. Some of these national economies are growing fast and becoming increasingly attractive targets for foreign investments. At the same time, other nations are poor, their economies still extremely undeveloped, and they are strongly dependent on international aid. This factor makes Chinese actions concerning the use of the rivers appear different depending on the observer. Particularly in the case of Thailand, China shares common interests in the areas of developing trade, transportation, and energy production. Thailand, with its strong industrial development, can import energy from China once the great hydropower plants are completed on Lancang. Vietnam is also aiming in the same direction in terms of developing the national economy and industry. In addition China and Myanmar share a somewhat similar approach to keeping their internal politics their own business, China has also supported Myanmar economically, which makes their relations quite close. On the other side lie poor countries such as Cambodia and Lao PDR, which are economically dependent on their stronger neighbours, and often forced to act according to their wishes. Even with good reason, there seems to be very little will or courage in these countries to openly criticize Chinese actions openly on the trans-boundary rivers.

The easiest way to cooperate in the international field is when the countries in question share common economic and other interests. However, this kind of cooperation is often narrowly focused, and is not necessarily sustainable. Even in the fast developing Thai society, the actions aiming exclusively at economic benefit cannot avoid questioning. The power of civil society, and especially NGOs is slowly growing in all these countries, and even if official statements support China's plans, the non-official groups acts strive to keep sustainability and environmental values on the agenda, hence opposing some Chinese plans.

The problems with China's international cooperation in water matters are nevertheless not unique. In fact, similar questions arise when dealing with developing economies and scarce water resources. For example, in the Ganga-Brahmaputra basin where India, Bangladesh, and Nepal are major players, it has been identified as a problem that the water experts in one country have only a very limited access to information from the other co-basin countries. It has been suggested that an operational mechanism for wider sharing of meteorological, hydrological, economic, and environmental information would alleviate this kind of situation (Biswas et al., 1998). But as sharing information and data is often considered to be sensitive, it is not an easy task. In the case of Mekong, the recent agreement China has signed with the countries in the lower reaches to inform them about rising water levels is a positive step in this direction. However, it is not yet enough to ensure long-term sustainable development in the region.

China and the Mekong River Commission (MRC)

It is quite significant that China, as well as Myanmar, did not join the MRC when the Mekong River Committee was reorganised in 1995. The Commission is not only a cooperative programme but also a part of the governance structure in each of its four member countries through the National Mekong Committees. At the moment, China is a so-called dialogue member of the MRC. The status means that Chinese representatives do participate in MRC meetings but not in the decision making, neither do they need to commit themselves to any agreements. However, cooperation seems to be slowly moving to a more practical level with the first agreements on information sharing of hydrological data on the Lancang-Mekong. Under the agreement, China will offer data on water levels of the Mekong as well as the rainfall in the river valley to the MRC every day during the flood season from 15 June to 15 October, until 2006.

In practice, the Chinese representatives taking part in the meetings chiefly come from the Yunnan provincial level, and not from the central government. The real status of cooperation with MRC is difficult to discern as different levels of Chinese goverment seem to have different approaches to it.

The reason why China is so far not a member of the Commission seems to be quite complicated and unclear. The question whether the country was really invited to join the Commission during its establishment raises different answers. Some Chinese specialists claim that the country was never even invited. However, according to many opinions from both China and outside its borders, China itself did not want to join. Being a member would restrict China too much in executing its plans to develop Lancang. Another difficulty for China seems to be that, as a full member of the Commission, it should be sharing information much more widely than it does today.

Whatever the original reason for not being a member of the MRC, the important question is whether China will join the Commission in the future. There seem to be quite fundamental challenges for the membership, concerning not only the Chinese side but also the MRC countries and the general development of the river basin area.

GREATER MEKONG SUBREGION (GMS) PROGRAMME

The GMS Programme was established in 1992 to facilitate sustainable economic growth and to improve the living standards of the people of the GMS countries. All six riparian countries, China, Myanmar, Thailand, Lao PDR, Cambodia, and Vietnam, are members of this programme which is supported by the Asian Development Bank (ADB). The GMS region covers 2.3 million km², and has a population of 230 million. In the past ten years, the programme has attracted US$ 1 billion in investment for close to 100 projects.

The programme covers seven different sectors: transportation, energy, tourism, trade and investment, telecommunications, human resource development, and environment (ADB, 1998). The GMS programme reportedly also includes replacing drug crops, human resources training, and hydro-electric dams. The most significant plans under the GMS are the 1855-km Kunming-Bangkok highway and the Pan-Asian railway that links Kunming and either Singapore, Thailand, or Myanmar (Wang, 2002). Establishing free trade zones between the countries is also on the agenda of the GMS. For China, increasing free trade with Thailand is of particular interest (Zhang Wei, 2002).

Participating in the GMS programme is clearly more agreeable for China than participating in the MRC. The GMS programme is one of the five key areas of economic cooperation between China and the ASEAN countries. It is an important corridor linking China with South-east Asia and is thought to serve as a model for any potential free trade zone between China and the ASEAN in the future.

China's most important aim in international cooperation seems to be of of infrastructure development for economic growth. The country has taken an active role in the activities of the GMS. The aims of the GMS are somewhat common to all participating countries. During the interviews, many commented that China would actually want to join the MRC, so that the discussion focused not exclusively on environmental issues but also on other aspects. This is important since economic matters cannot be separated from the environmentally sound use of natural resources.

WHAT KINDS OF FACTORS LIE BEHIND CHINA'S COOPERATION IN INTERNATIONAL WATER ISSUES SO FAR?

Certain assumptions can be made for the factors influencing the way that China cooperates. The basic assumption is that the level of China's international cooperation in water resources management was relatively low. This appears to be a correct assumption.

The structures of both the Chinese society and politics affect its international activities. Strong central administration, one-party system, and a tendency towards keeping internal matters as the nation's own business make the borders tighter than in some societies where cooperation is more evident. The structure and working methods involved in water management makes it difficult to practise integrated management. The duties are divided among different ministries and bureaus so that there is hardly any information sharing among them. As a consequence, overlaps exist, as the basic information is collected and often the same work is done in many of these bodies. To fulfil the aims set for them, the offices also compete with each other, which reduces further the will to share information. In the middle and lower levels of governance, contradictions exist in the position of the bureaus. They need to respond both to the next level in their field and to the general local governance. These very basic challenges in turn have their impacts on international cooperation.

For historical reasons—the tradition of great empires and position as a superpower—there remains a view within China that it is distinctly above its neighbours. Its huge physical size, and fast industrial development in recent years, differentiates it from its neighbours. The turbulent history of the entire region of South and South-east Asia, including political disputes and wars in the recent past, makes it even harder to reach a comprehensive and fair level of cooperation.

There are some important challenges that China faces today. Environmental degradation is one such challenge, that needs urgent attention. The same problems as mentioned before, a non-optimal management structure and lack of funding also occur in the environmental field.

The importance of projects that support internal development, such as energy and transportation sectors, must not be underestimated. Water resources have a significant potential from this point of view. This approach gives the projects on trans-boundary rivers a clear-cut and strong justification for China, even if the impacts on the other riparian countries are negative.

The politically very sensitive question, which occurs with almost all the international rivers in the area of south-west China, is that of Tibet. Almost all the great international Himalayan rivers originate from the Qinghai-Tibet Plateau, and flow south or south-east to the Yunnan province and then to other riparian countries. Developing these rivers raises the question of the indigenous people of Tibet. The livelihood of the local people, environmental impacts and China's tendency to use natural resources for its own purposes are difficult issues which cannot be avoided in open discussions of the great Himalayan rivers.

A positive sign of the willingness to cooperate was found during the interviews at the ground level. The importance of working together was understood within the local administrations responsible for the Chinese side of the trans-boundary rivers. A clear will exists to act responsibly so that the impacts on the other riparian countries would be modest. However, these officials have no real opportunities to work towards these aims. The main reason is the serious lack of resources. Funding is often not available even to buy the technology and equipment required for control and observation. Building the capacity of the officials, which is necessary for keeping abreast of the latest technological developments has not been possible. It is not realistic to ask these officers to practise comprehensive international cooperation when they do not have the opportunity to take the necessary actions for their very own prefecture. All these resource questions feed back, to the priorities of the Chinese government.

The lack of information concerning the Chinese side of the river basins has been presented as one of the reasons hindering China's entering into new international agreements. It is true that little information is available as many of these rivers are in remote areas, and collecting data would require considerable funding and human resources. Work to obtain more comprehensive information has been started, and for the Lancang-Mekong there should be more hydrological data available in the future. The real impact of the lack of information can still be questioned, particularly when comparing countries in a similar position. It is unlikely that poor countries such as Lao PDR and Cambodia can obtain more data on their areas than China has today. All these add to the complexities of managing these international rivers.

The binding nature of international treaties is a factor that does have an important meaning for China. To get China involved in such treaties, it might help to make the treaties more flexible. On the other hand, the purposes for which the agreements are needed are often so sensitive that an agreement must be a compromise among the countries involved. All the players have their own needs, and a common aim that would benefit them all equally is hard to set. Thus, the rules that the participants follow must be accurately defined.

Is China Willing to Cooperate More in the Future?

Among the most essential factors in cooperation is the benefit to be gained. If China's interaction is seen to be minimal, the logical outcome is that there is a lack of adequate benefit for China.

As China opens futher, it would seem natural that the willingness to cooperate internationally in any field will increase. However, nothing can be taken for granted.

During the interviews conducted for this analysis, the general picture was that a willingness exists in China for more international cooperation. However, according to some specialists, certain driving forces must be in place before any serious development on this front can take place. First, there needs to be enough information, hydrological data for example, about the Chinese stretches of the rivers. For example, in the case of the Lancang, China claims that they do not have accurate data of the river yet, but have recently started projects to collect it (He, 2002).

Second, the form of cooperation must be planned so that China does not feel being manipulated by the others. It matters too what fields are included in the international agreements. China wants the cooperation to cover more than the environmental side of water use, which can be seen for example in its participation in the GMS programme while staying out of the MRC. According to the most sceptical opinion, China might want to participate in such organizations as the MRC in the future, after it has completed all its plans for the rivers.

It can nevertheless be argued that the challenges above are met always when talking about international and integrated water resources management. Still, some positive results have been reached elsewhere. What makes China so different from the others? China behaves similarly to other regional or global great powers, which more or less ignore the needs of their neighbouring countries in issues important to themselves. On the other hand, nobody knows if these driving forces that China is claiming will ever be completed. In this light, sitting and waiting for something to happen will not improve the situation.

Chinese specialists have made some proposals to increase cooperation and sustainable management among the riparian countries. As problems concerning China's cooperation with the neighbouring countries, they mention among others the lack of organized or systematic study concerning the international rivers, and the lack of effort to set-up an organization for development and management (He et al., 2001). The suggestion is to increase the level of both research and joint work among the riparian countries.

WILL CHINA JOIN THE MEKONG RIVER COMMISSION?

The question of China's possible membership of the MRC in the future has two aspects: the point of view of China and that of the neighbouring countries. For China, the restrictions on the use of the water that obviously would follow the membership are difficult to approve, as there is a great pressure in the country to develop the almost untapped rivers. If China joins the commission, it is quite clear that it could not fully complete its plans for the Lancang.

It has also been said that China is not willing to join the Commission because it is currently a cooperating body, where all nations have an equal position. According to this thinking, a new formulation of the Commission may be required before China could agree to join.

It is quite surprising that, despite the common belief that the MRC countries wish China to join, there has been speculation over whether these four countries really do want their huge neighbour to become a member. As for joining the Commission, China's membership would probably somehow dominate the actions of the Commission and may change its role.

One difficulty is to define which governmental level in China would be the real actor to promote more cooperation. At the moment, as China has the position of a dialogue member, it is mostly the Yunnan province that participates in discussions and meetings. In the Commission, the parties are nevertheless nations, and thus it should also be the central government of China. As the provinces in China are in quite a strong position for making their own decisions, it might become complicated to define the decision-making process internally in the country. In addition, the most southern Chinese prefecture along the Mekong, Xishuangbanna, has a special position as an autonomous prefecture. The main reason for this is that the area has the greatest number of ethnic minorities in the country. The prefecture has its own objectives in developing the area, which sometimes are not shared by the central or even the provincial government. As regards Lancang-Mekong, Xishuangbanna prefecture has presented a wish to the central government for China to join the MRC as they see many benefits in stronger international cooperation (Li Fengling, 2002).

On the other hand, if China joins the Commission, China would want it to cover more than only environmental questions, and to be enlarged to include economic and trade cooperation. At the moment, there exists another form of cooperation for trade and transportation in the area, the Greater Mekong Subregional (GMS) Programme. If China is to become a member of the MRC, the Commission's interests might have to move towards a more consistent approach than that of the current MRC.

WHICH DRIVING FORCES MIGHT INCREASE CHINA'S WILLINGNESS TO
COOPERATE—WHAT COULD ACT AS A MOTIVE FOR CHINA?

Finding the positive driving forces for China is an important step in
encouraging the countries to closer interactions. The question is: what
could act as a motive for China to cooperate more in future use of water
resources. The attention in this section is focused mainly on the economic
factors related to South-east Asia, China's policy of opening up to the
outside world, and increasing participation in the international community,
as well as public participation.

The potential for trade that lies particularly in the developing economies
of South-east Asia is certainly one factor that affects Chinese decision
making. China has pressures to both sell goods to the riparian countries
of the Mekong and to import such necessities as grain from these areas.
China's relations with the ASEAN countries seem to have become closer
during the last years. As far as South Asia is concerned, the economic
benefits are not so clear, which might partly be due to the sensitive
balance China and India are maintaining between them as the two great
powers in the area.

The concern for the degradation of the nature and biodiversity in many
of the great Himalayan rivers, or at least the high risk to the natural
environment that any plans to develop them would include, is a topic that
particularly NGOs and international environmentally focused organizations
try to get into decision making. However, although environmental values
have been given more weight in China during the recent years, they still
come only after economic development. This approach concerns acting in
both internal and international matters.

In summary, the direct economic benefits seem to get the highest
position among the various factors affecting China's international
cooperation in water resources management. There are other important
matters, but they can only colour the picture. Before criticizing China for
this approach, it must be remembered that most of the other nations also
follow the rules of market economy, and thus seek economical benefits.

Conclusions

China's great challenges in the water sector have been studied broadly in
recent years. There seems to be a relatively common understanding of the
importance of water resources management in the country, and much
effort is put into projects to improve the situation. However, one aspect
of water management has not yet been observed—the international

dimension and significance of China's activities on the trans-boundary rivers.

Even though many efforts are under way to solve China's water problems, the future is unclear as the country's focus remains strongly on economic development, and the sustainable use of natural resources remains in second position.

In Chinese literature, the use of integrated water resources management concept is presented more widely than it seems to be in reality. It is claimed that the country is moving towards comprehensive water management and taking the entire river basin into account when planning the allocation of water resources (He et al., 2001). For example, in his paper 'Integrated Water Resources Management: Theoretical Exploration and Practices in China', Shen (2003) claims China is pursuing integrated water resources management.

After the serious floods of 1998, institutional modifications were carried out, and the newly formed Ministry of Water Resources did restructure its policies of water resources management. Water rights should now be allocated among the different forms of use and different regions, following certain principles; water pricing forms an important part of the fair allocation. According to Shen, water resource-bearing capacity and water-related-environment-bearing capacity should be taken into account, and environmental water requirement be satisfied first, before economic requirements. The new Water Law, put into force in 2002, is supposed to resolve the problems and insufficiencies of the older law, and strengthen IWRM practices in China.

Unfortunately, the messages from the field reveal that integrated management only happens in theory. However good the plans are and however much the government promotes the concept, the implementation of integrated management remains far from reality, at both the internal and international levels. Even Shen admits, when discussing environmental degradation and the needed flood control capacity that has not been reached, that there is a lack of integrated planning and management. It could be said that the term 'integrated water resources management' remains not commonly understood or used in China.

One of the major problems in implementing real cooperation is said to be the language barrier. Because of financing problems in translating any publications, Chinese research can only be read in Chinese language. According to this proposition, the other riparian countries are still not fluent in Chinese to read the results of Chinese research. However, it would seem strange that in the neighbouring countries, where ethnic

Chinese exists, people with adequate language skills cannot be found. The other question is whether the publications are available outsite the Chinese borders, which seems to be the case at least with the Lancang-Mekong environmental impact assessment. However, as in all the information exchange in South and South-east Asia, the lack of funding and resources plays a role in sharing hydrological and environmental data. However, even more important than funding is the real will for open discussion and information sharing, among the countries, which is still missing.

Certainly, other points of view besides those of the government exist. People working at local and ground level seem to be interested in sustainable use of the water resources in cooperation with the neighbouring countries. Unfortunately, there are no real possibilities to do it. The lack of resources is serious, as there are not even enough funds to buy the technology and equipment, or to train officials. The lack of funding is an understandable explanation for minimal cooperation, and is probably also a real reason.

The question of the will for international cooperation seems to have two entirely different sides—the knowledge and understanding on the other side and the existing state of reality. The problem is not the ability of Chinese researchers in the field of water to understand the challenges and cures of the trans-boundary questions. The expertise in the country is good, but the focus of water management is in many cases still too narrow, as each institute or department concentrates mainly on its own specific area of interest.

It seems that for China it is important to cooperate internationally in matters concerning economic development. However, when it comes to more contentious topics such as the use of water resources, the willingness to share the country's benefits for the sake of cooperation seems to diminish remarkably. On the other hand, it is natural for any nation to prioritize its own welfare, and China is just doing the same. The point is to find rules for the game that enable all players in the game, both small and big, to win. There must be something to be gained for the one that enters into an agreement. Therefore, in this case, one cannot expect China to cooperate only to benefit its neighbouring countries; world simply does not operate that way.

References

Baldizzone, T. and G. Baldizzone, *Tales from the River—Brahmaputra*, Boston: Shanbhala Pulications Inc., 1998.

BBC News Online, 2004. Thai groups battle new China dam. Tom Butler, January 19, 2004. Available online at http://news.bbc.co.uk

Biswas, A.K., M. Nakayama and J. Uitto, 'Sustainable development in the Ganges-Brahmaputra basin', in *Global Environmental Change*, Vol. 8, No. 3, 1998, Elesevier Svience Ltd, Great Britain, pp. 275–7.

Campbell, I., 'The Mekong River Basin and Integrated Water Resources Management', in *Integrated Water Resources Management in South and Southeast Asia*, A.K. Biswas, O. Varis, and C. Tortajada (eds), New Delhi: Oxford University Press, 2004.

Cao Meng Liang and Dong Xue Jun, *Personal communication*, Jinhong, China: Xishuangbanna Forest Bureau, 2002.

Chen, L., 'Xiaowan Dam, a reservoir for progress', A paper presented at the seminar, 'Our Mekong: A Vision Amid Globalization', Chiang Rai, Thailand: sponsored by Inter Press Service (IPS) Asia-Pacific, 2002.

Feng, Y., *Personal communication*, Kunming, China: Yunnan University, 2002.

He, D., *Personal communication*, Kunming, China: Yunnan University, 2002.

He, D., C. Liu, and Z. Yang, 'The Study on the sustainability of the international rivers in China', in He Daming, Zhang Guoyou, and Hsiang-te Kung (eds), *Towards Cooperative Utilization and Coordinated Management of International Rivers*. Tokyo: The United Nations University, and New York: Science Press, 2001, pp. 9–21.

Institute of Geography, the Chinese Academy of Sciences, *The Natural Features of China*, Beijing: China Pictorial Publishing House, 1992.

Kajander, T., *Water Resources, Large Dams and Hydropower in Asia*, Espoo: Helsinki University of Technology, 2001, *http://www.water.hut.fi/wr/research/glob/publications/kajander/alku_v.html*

Kummu M., Sarkkula J., Varis O., 2004. Sedimentation and Mekong Upstream Development: Impacts to the Lower Mekong Basin. Conference paper, the IAG Yangtze Fluvial Conference in Shanghai, China, on June 2004.

Li, Fengling, Jinhong, China: *Personal communication*, Xishuangbanna Environment Protection Bureau, 2002.

Li, Ging, *Personal communication*, Jinhony, China: Xishuangbanna Transportation Bureau, 2002.

Liu, Heng and Haixing Tang, 'The Comparative Analysis of the Hydrological Characteristics in the Lancang-Mekong River Basin', in He Daming, Zhang Guoyou, and Hsiang-te Kung (eds), *Towards Cooperative Utilization and Coordinated Management of International Rivers*, Tokyo: The United Nations University, and New York: Science Press, 2001, pp. 9–21.

Liu, Tianchou, 'The hydrological characteristics of the Yarlung Zangbo River' in, He Daming, Zhang Guoyou, and Hsiang-te Kung (eds), *Towards Cooperative*

Utilization and Coordinated Management of International Rivers, Tokyo: The United Nations University, and New York: Science Press, 2001, pp. 104–11.

Makkonen, K., *An Analysis of China's Role in Integrated Water Resources Management (IWRM) in South and South-east Asia*, Espoo: Helsinki University of Technology, 2003, *http://www.water.hut.fi/wr/research/glob/publications/ makkonen/alku_v.html*

McCormack, G., 'Water Margins: Development and Sustainability in China', Australian Mekong Resource Centre (AMRC), Working Paper Series No. 2, 2000.

New York Times, 2004a. In surprise move, Wen cites concern for environment. Jim Yardley, April 8, 2004. Available online at http://www.iht.com

New York Times, 2004b. Dam Project Threatens Chinese 'Grand Canyon'. Jim Yardley, March 10, 2004. Available online at http: query.nytimes.com

Paopletto, G. and J. Uitto, 'The Salween River: is international development possible?' in *Asia Pacific Viewpoint*, Vol. 37, No. 3, 1996, Blackwell Publishers, Oxford/Cambridge.

Shen, D., 'Integrated Water Resources Management: Theoretical Exploration and Practices in China', unpublished manuscript, Department of Water Resources, China Institute of Water Resources and Hydropower Research, 2003.

'TA for the Strategic environmental Framework for the Greater Mekong Subregion', Manila: Asian Development Bank, 1998.

Uitto, J., *Personal communication*, New York: UNDP-GEF, 2002.

Wang, W., *Personal communication*, Kunming, China: Yunnan Research and Coordination Office for Lancang-Mekong Subregional Cooperation, 2002.

Xinhua, 'China to build huge power station on Lancang-Mekong river', *China Daily*, 2002a.

Yu, Xiaogang, 'Change flows down the Mekong', *Bangkok Post*, 2002, *http://www.bangkokpost.com*

Zhang, Wei, *Personal communication*, Beijing, China: The China International Centre for Economic and Technical Exhanges (CICETE), Ministry of Foreign Trade and Economic Cooperation, 2002.

11

INSTITUTIONS FOR INTEGRATED WATER RESOURCES MANAGEMENT IN LATIN AMERICA: LESSONS FOR ASIA

Cecilia Tortajada

Introduction

The global interest in environmental and social issues has increased since the convening of the United Nations Conference on the Human Environment in Stockholm in 1972. When the Stockholm Conference was held, only eleven countries had appropriate institutional arrangements to manage environmental issues. Exactly twenty years later, when the United Nations convened its next global Conference on the Environment and Development in Rio de Janeiro, nearly all the member countries of this organization had established institutional mechanisms to manage their environmental issues at the national level. One of the aspects that contributed to this development was the individual and collective recognition of the countries on the increasing degradation of their natural resources and its inevitable long-term adverse linkages with their future economic development, poverty alleviation, and overall quality of life of their populations. This awareness prompted political attention towards social and environmental issues at the national and international levels, which resulted in several worldwide gatherings to analyse development and environment-related issues. Unfortunately, more than ten years after the Rio meeting, continuing environmental degradation, including water resources, suggests that many developed and developing countries have still not managed to formulate and/or implement proper public policies which can successfully address environmental issues within their overall economic and social frameworks.

In the area of water resources, the dominant trend of the 1970s and 1980s was on the construction of infrastructures. During the early 1990s, the focus shifted to the management of water resources, but still largely on sectoral lines. By the end of the 1990s, integrated water resources management (IWRM) had become the main trend, with a broader multi-sectoral approach, and consideration of social and environmental issues. Stakeholder participation became an important aspect of water planning and management. It was recognized that many of the activities related to IWRM like demand management, efficient and equitable allocation of water among uses and between users, environmental and social impacts, stakeholders participation, etc., had little to gain from a paradigm which focused mainly on the construction of water infrastructures. It was further recognized that in order to achieve IWRM, appropriate institutions would have to be developed, with adequate managerial and technical expertise and financial resources. The World Bank (1993) and the Asian and Inter-American Development Banks changed the focus of their policies from development of water resources to their holistic management. The emphasis of these international agencies shifted from sub-sectoral and project-based development projects to issues like demand management, private sector involvement, water pricing, environmental conservation, social participation, and river basin management. In other words, the new policies were expected to take into account social, environmental, and economic aspects, which would result into more effective regulations, incentives, investment plans, and environmental protection.

To a certain extent, the shift in the policies of the World Bank and the regional development banks were partly because of the continued evolution of the concepts of sustainable development, IWRM, integrated river basin development, and advances in knowledge-base and technology. Additionally, it was further recognized that investments supported by international funding agencies had historically faced shortcomings during the implementation of water-related projects mainly due to weaknesses in national policies and institutions (World Bank, 1993). It was realized that infrastructural projects do not necessarily improve social or economic conditions by themselves, but it is proper planning and investment, as well as technical, financial, and social support, which are decisive for their overall success.

These concepts (sustainable development, IWRM [GWP, 2000], and river basin management) have proved to be appealing, and have received widespread global acceptance from national and international institutions in recent years, including from Asian and Latin American countries. However, their effective incorporation and implementation have proved to be extremely difficult, irrespective of the country concerned. There are many reasons for these shortcomings, among which are inadequate institutional and legal frameworks for integrating environment and development-related issues, highly centralized decision making, absence of political will to change the status quo, lack of adequate number of qualified and trained personnel, non-availability of financial resources on a timely basis, and the sheer difficulty to look at all water-related issues in an integrated manner, specially as the water sector affects many other sectors (agriculture, energy, industry, environment, tourism, etc.) (Dourojeanni, 2000) and in turn is affected by the others. In addition, processes like decentralization and privatization have yet to produce results that are expected from them in most countries: more efficient water distribution services, encouragements of local initiatives, active participation of water users, generation of new investments funds, and involvement of diversified actors.

Integrated Water Resources Management

Development and environment-related problems have remained largely a disciplinary-based exercise in all sectors in most Asian and Latin American countries, and the water sector is not an exception. The environmental issues are often analysed without consideration of their economic, social, and cultural interlinkages. The dominance of this limited approach in the formulation of environmental policies has contributed to the continued acceleration of environmental degradation, in spite of the widespread political and social interest in its abatement (Meppem and Bourke, 1999). So far as water resources is concerned, solutions to specific water issues cannot always be found within exclusively the water sector. Answers often have to be formulated within an overall inter-sectoral framework, on which water professionals may have limited say or control. All major water issues are interlinked with many other development-related issues, which means that both the problems and their possible solutions do not depend, and thus should not rely, exclusively on any one discipline, institution or sector. This recognition has further contributed to the widespread acceptance of the concept of IWRM, which considers water management on a multi-sectoral and multidisciplinary basis.

In most of the developing world, it is being gradually recognized that the increasing complexities of the problems associated with efficient water resources management, and accelerated societal interest in water issues, necessitate a broader approach, which cannot be provided by engineers and/or economists alone. Irrespective of the theory, however, the reality is that the traditional institutions are still basically sectoral, where most of the sectors tend to focus on activities related to one specific use of water and not to their multiple use or management at appropriate regional levels (Dourojeanni, 2001a).

In addition, it is important to remember that individuals live and operate in a world of institutions, where opportunities and prospects depend on which institutions exist, how effectively they function, and how responsive they are to the changing needs and requirements of the society. However, many planning and execution programmes developed by water institutions in most of the developing world have not achieved the degree of integration required to take appropriate and timely actions with reference to issues like water resources management, consideration of economic, social and environmental impacts, human resources development, governmental and non-governmental participation, etc.

If IWRM is to be a goal, it is more likely to be achieved if management units are defined. One option could be the management at the river basin level, which is potentially one promising institutional alternative, if, and only if, it is developed within a framework of integrated water resources planning and management, and in consideration of economic and social development, and environmental protection.

Management activities at the river basin level may vary depending upon the policy objectives, which include decentralization, public participation, infrastructural development, environmental conservation, rural development, land and water management, etc. The integrative natural resource management at the river basin level can be based on either natural resources in general, or, as is the current case for Latin America, with focus almost exclusively on water resources (Dourojeanni, 2001b). In this case, it is desirable that the interests of the different users, whose economic and social activities depend on the river basin, are represented in the policy making of the region so that no conflicts arise among them and that their activities are not affected adversely.

The river basin bodies are expected to plan and act in response to specific regional needs, promote the formulation of region-specific strategies and assist in their implementation, take actions for rational and efficient use of resources to improve the lifestyles of the local people, and simultaneously protect the environment (ECLAC, 1997). Depending upon the roles, jurisdiction, and authority of the river basin organizations, they can be classified as authorities (with statutory power to promote and enforce changes), entities (intermediate and varying in power) and commissions or committees (with advisory or monitoring roles, but no decision-making power) (Barrow, 1998). The type of river basin organizations that are needed depends on the policies of the countries, objectives that are to be achieved, and management capacities available. The roles, responsibilities, and financial resources of the organizations, have to be defined within the existing legal and institutional frameworks of the specific countries concerned for a more efficient performance. With the recent decentralization drive in most Latin American countries, the roles of the water management institution at different levels should be clarified in order to ensure a better coordination between them. For example, river basin organizations have to be carefully incorporated into the administrative structures of the countries to promote cooperation among institutions at the central, state, regional, and municipal levels, and to avoid duplication of activities, inefficiency, and lack of resources, both human and financial.

River basin organizations have been created all over the world. However, the focus of the present paper is the performances of the river basin organizations in Latin America only, with special emphasis in Mexico.

River Basin Management in Latin America

Interest in many Latin American countries in river basin organizations is not new: it has existed for over half a century. It originated as an attempt to replicate the experiences of the Tennessee Valley Authority (TVA) of the United States. The approach to use river basins as management units at that time in the region could be partly explained by the then prevailing national policies that favoured a regional planning approach which could create new development poles so as to ease the pressures on the existing urban and industrial centres. The TVA model was generally viewed as a possible way to promote development and accelerated industrialization. Some regional development corporations were established on the basis of river basins, like several commissions in Mexico towards the end of the 1940s and the beginning of the 1950s (Barkin and King, 1986; Garcia, 1999).

The French and the British experiences on river basin management were also noted with considerable interest in many Latin American countries. However, these European experiences were not critically or objectively analysed to determine to what extent river basins could be the most optimal units for water management, and their relevance and potential replicability under the differing physical, environmental, social, economic, legal, and institutional conditions of Latin America. This acceleration of interest in river basins have resulted in the formation of several regional institutions like the Latin American Network for Watershed Management, and regional networks such as the Latin American Network of Basins Organizations (LANBO). At the regional level, many bilateral and multilateral commissions were created, like the Bermejo and Pilcomayo Commissions, Inter-Governmental Plata River Commission, the Amazon Cooperation Treaty, etc. (Biswas et al., 1999). At the national levels, there has also been a general trend towards the establishment of river basin organizations, although in many cases, a centralised top-down management structure primarily along sectoral lines continues to be in place.

Geographical, economic, political, social, legal, and institutional considerations have been the main drivers for the improvement of the water management practices in different countries. Furthermore, it should be noted that Latin America is neither a homogeneous region, nor are all the countries at the same level and stage of socio-economic development or have similar management capacities and technical expertise. Thus, any single institutional model is highly unlikely to be equally applicable for efficient water resources management for all the Latin American countries. In spite of this, the region as a whole appears to be moving towards river basin management, although one cannot prejudge a priori the results of this shift at present. The concept of idealized situations where things were expected to be 'as they should be', needs to give way to a more pragmatic approach, where river basin management activities must conform to demands and constraints imposed by real situations, where things are 'as they are' (Garcia, 2001).

Some Latin American countries have developed, or are in the process of developing, institutional frameworks which focus on the management of their water resources at the regional level by creating river basin organizations. Water laws have also been modified due to the urgent need to improve water management practices which can match the growing competition between its multiple uses and users, especially due to the increasing demands for water in urban areas, irrigated agriculture, hydropower generation, and environmental protection, and simultaneously manage its quality, in addition to the desire to encourage the participation of the private sector in supplying water-related services, and enable users to administer water themselves (Marañón-Pimentel and Wester, 2000; Dourojeanni, 2000). The present thinking is that extensive participation of the private sector and water users will significantly improve the performances of the existing water institutions. Finally, another reason is the need to formulate a more modern legislative framework which could contribute to control floods and droughts, the over-exploitation of aquifers, and the increasing deterioration of water quality in all the Latin American countries. All these expectations are to be fulfilled through what is being extensively promoted as IWRM.

So far at least in terms of legislative frameworks, only Brazil and Mexico in Latin America have water laws which promote this approach. The Inter-American Development Bank has already financed, or is likely to finance, more than twenty projects at the basin-management level in countries like Argentina, Bolivia, Brazil, Chile, Colombia, Costa Rica Dominican Republic, Ecuador, El Salvador, Guatemala, Haiti, Honduras, Jamaica, Nicaragua, Panama, Peru, and Venezuela. Even though the river basin concept in the region was started mainly to deal with a single resource, water, the current approach is expected to integrate economic, social, and environmental issues simultaneously (Garcia, 1999; 2001). To what extent such an integrated management is realistically possible when the management unit is primarily chosen on the basis of water, is basically unknown at present. This is an especially pertinent issue, since at the national or regional levels, countries have institutions that manage mostly separately, water, environment, agriculture, and economic and social affairs. How all these and other associated activities can be coordinated and integrated within the context of a river basin is still to be seen. In the final analysis, efficient administration of the natural resources, water being only one of them, at the river basin levels is likely to depend primarily upon the extent of decision making power and financial autonomy that would be delegated to such authorities (Solanes and Getches, 1998).

Experiences from integrated management of river basins in Latin America in general, and specific case studies from Argentina, Brazil, Chile, Colombia, Mexico, Peru, and Venezuela, indicate that the units for optimal water management need to be further studied and objectively reviewed before any definitive conclusions can be drawn (Biswas et al., 1999; Biswas and Tortajada, 2001). It may be quite likely that the units for water management may not necessarily be in terms of river basin units for a variety of technical, political, institutional, legal, social and cultural reasons. In some cases, the units could even be best managed according to administrative regions. In addition, the most appropriate units for water management may even vary from one Latin American country to another because of the wide diversity of conditions. This is an area that needs further serious investigations.

As noted earlier, Brazil and Mexico are the only two Latin American countries where river basin organizations are legally mandated. Examples of river basin management practices from Mexico are discussed next, which illustrate the present status of the application of the concept.

MEXICAN EXPERIENCES

Mexico is a country with 1.47 per cent of the world's surface, but it occupies the fourth place in terms of biodiversity. It has 10 per cent of all known species, and a wide array of unique species not found anywhere else. However, a striking characteristic of the country's economic structure is the fact that a significant proportion of the territory (64 per cent) considered as having low or very low quality of natural resources, is the area in which more than one-third of the population lives, and where 40 per cent of the production value is generated (see Figure 11.1). This disparity has meant that in some regions the resources are over-exploited and in other regions, natural resources, including water, are underutilized.

An interesting issue is the imbalance of water resources among regions, which are constrained by the following three mismatches. First, annual rainfall is concentrated in four months and a large part of it is lost. Second, most water consumption occurs in the north, while rainfall and water reserves are more abundant in the south. Third, 80 per cent of the water storage is below 500 m in elevation, whereas 75 per cent of the population lives above this elevation. Paradoxically, despite the abundance of water resources in the south, this area contains the states having the highest percentage of households living in poverty and the lowest percentage having access to clean water (40.2 per cent) and sanitation (25.9 per cent), with attendant health risks associated to these deficits (OECD, 2003). According to OECD's study on Territorial Reviews in Mexico (2003), one of the main reasons for Mexico's spatially differentiated growth patterns is that successive Mexican governments have traditionally maintained ambiguous and inconsistent approaches towards regional development while emphasizing mainly sectoral policies defined at the federal level, has not addressed properly considerations of their explicit or implicit regional impacts.

In Mexico, water demands have historically been met through the construction of new infrastructures without consideration of either the validity of the consumption patterns or the efficiency of the existing water use and management practices. Therefore, historically, adequate considerations have not been given as to how water problems can be solved through more efficient water management practices, including demand management.

Towards the end of the decade of the 1940s, the Mexican government embarked on large-scale water-based regional development programmes. The first group of river basin commissions (Papaloapan, Grijalba, Fuerte, Tepalcatepec, and Balsas) were established to coordinate the activities of the different ministries working in the several states of the region where

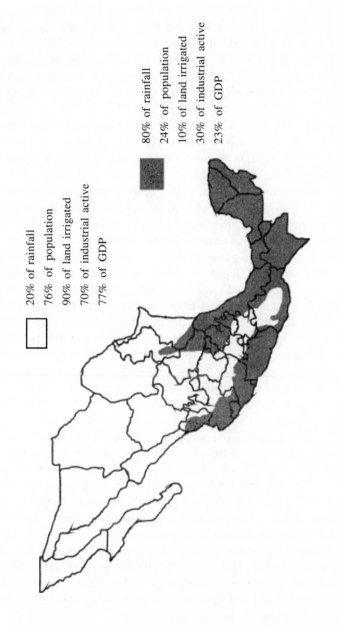

20% of rainfall
76% of population
90% of land irrigated
70% of industrial active
77% of GDP

80% of rainfall
24% of population
10% of land irrigated
30% of industrial active
23% of GDP

Figure 11.1: Spatial distribution of water availability, population and economic activities

the programmes were initiated. The funds for this large-scale investment programmes were allocated to the commissions, as well as a wide spectrum of responsibilities. The commissions had no autonomy and were designed to be dependent on the federal authority responsible for water. These river commissions proved to be reasonably effective instruments for the implementation of the regional policies of the country. Their tasks included not only financial matters at the regional levels, but also planning and coordination activities which were earlier the responsibility of ministries and governments at the state level. The main investment programmes of the commissions included initiatives like water resources development, construction of irrigation projects, and programmes on flood control and hydropower generation. However, they were also responsible for expenditures on roads, schools, public health issues, and so on, for the regions. The power, and thus efficiency, of these commissions depended on the priority the specific region had at the national level. Their main weakness, however, was that they were created as coordinating and advisory units, with no power to force any public or private institutions to comply with any legislations, or to question any unwarranted environmental, social, and economic impacts (Barkin and King, 1986).

Later, during the 1960s and 1970s, large-scale rehabilitation projects were undertaken to increase the productivity of the existing irrigation districts. Plans for large-scale inter-basin transfers (Cutzamala and Lerma-Balsas systems) were developed to expand the irrigated areas in the north-west part of the country, as well as to ensure a source for future water supply to the Mexico City Metropolitan Area (Tortajada, 2000a).

The apparent abundance of financial resources in the country during the late 1970s and early 1980s reinforced the notion that water could be supplied at any cost, even when the cost was not covered by the consumers who benefited from increasingly subsidized water services in the cities as well as in irrigated areas. However, the economic and financial crisis of the 1980s had a definite impact on water development. As the federal government faced serious budget constraints, most water investment programmes were reduced to a minimum, certainly far below of what was needed to meet increasing demands, and also in closing existing gaps in water services. Available resources for investment programmes were further reduced by increased federal subsidies for operating and maintaining water services in cities and irrigated areas. In addition, these subsidies were insufficient, which further contributed to the progressive deterioration of water infrastructures (Biswas, 2003).

In 1989, the National Water Commission of Mexico (CNA) was created by a presidential decree as the sole federal authority to deal with water management as an autonomous agency. Initially, it was part of the Ministry of Agriculture and Water Resources. However, after the Earth Summit at Rio de Janeiro in June 1992, CNA was moved from the earlier Agriculture and Water Resources Ministry to the Ministry of Environment, Natural Resources, and Fisheries (SEMARNAP, 1997), which was renamed Ministry of Environment and Natural Resources (SEMARNAT) in 2001. In terms of legislation, in 2003, a new draft water law was presented, and accepted in April 2004 (Semarnat, 2004). It states that the water management of the country would still be based on basin organizations, and that the opinions and concerns of the users would still be channelled though general assemblies and basin councils, which will continue to depend on the federal authorities.

With the objective to manage water resources in a more coordinated way, Mexico has been divided into regions and sub-regions: there are thirteen administrative regions based on the hydrology of the country, and 102 sub-regions on the basis of political jurisdictions. Each sub-region includes a number of municipalities of the same state, so that regional programmes could be planned at the sub-regional level. At present, there are 314 hydrological basins, thirty seven hydrological regions, and thirteen administrative basins in the country (Biswas, 2003). About twenty-five river basin councils, out of the twenty-six that were planned, have already been established, but the vast majority are not functional yet (see Table 11.1).

At present, the basin councils are expected to manage water resources from integrated and regional perspectives, and involve water authorities at the federal, state, and municipal levels, as well as the various users. For operational purposes, the basin councils define four territorial levels: basin, sub-basin, micro-basin, and aquifers, where the bodies are respectively known as councils, commissions, committees, and groundwater technical committees (COTAS) (CNA, 2003) (Tables 11.2 and 11.3). By law, the river basin councils have to approve the river basin plans which, once integrated within the National Water Master Plan, become mandatory for the federal government, and indicative for the local and the state governments and water users. In these plans, bottlenecks and necessary actions and resources needs are to be identified and evaluated, and unrealistic or unfeasible situations are expected to be fed back into the regional planning process (Tortajada, 2000b).

Table 11.1: Basin councils as of 26 November 2002

Name	Date in which it was established	Administrative region
Baja California Sur	3 March 2000	I Peninsula de Baja California
Baja California	7 December 1999	I Peninsula de Baja California
Alto Noroeste	19 March 1999	II Noroeste
Rios Yaqui-Matape	30 August 2000	II Noroeste
Rio Mayo	30 August 2000	II Noroeste
Rios Fuerte y Sinaloa	10 December 1999	III Pacifico Norte
Rios Mocorito al Quelite	10 December 1999	III Pacifico Norte
Rios Presicio al San Pedro	15 June 2000	III Pacifico Norte
Rio Balsas	26 March 1999	IV Balsas
Costa de Guerrero	29 March 2000	V Pacifico Sur
Costa de Oaxaca	7 April 1999	V Pacifico Sur
Rio Bravo	21 January 1999	VI Rio Bravo
Nazas-Aguanaval	1 December 1998	VII Cuencas Centrales del Norte
Del Altiplano	23 November 1999	VII Cuencas Centrales del Norte
Lerma Chapala	28 January 1993	VIII Lerma-Santiago-Pacifico
Rio Santiago	14 July 1999	VIII Lerma-Santiago-Pacifico
Costas del Pacifico Centro	—	VIII Lerma-Santiago-Pacifico
Rios San Fernando-Soto La Marina	26 August 1999	IX Golfo Norte
Rio Panuco	26 August 1999	IX Golfo Norte
Rios Tuxpan al Jamapa	12 September 2000	X Golfo Centro
Rio Papaloapan	16 June 2000	X Golfo Centro
Rio Coatzacoalcos	16 June 2000	X Golfo Centro
Costa de Chiapas	26 January 2000	XI Frontera Sur
Grijalva-Usumacinta	11 August 2000	XI Frontera Sur
Peninsula de Yucatan	14 December 1999	XII Peninsula de Yucatan
Valle de Mexico	16 August 1995	XIII Valle de Mexico

Source: Statistics of Water Resources in Mexico, 2003, CNA, Mexico (in Spanish).

Table 11.2: Basin commissions as of 26 November 2002

Name	Date in which it was established	Administrative region
Rio Colorado	7 December 1999	I Peninsula de Baja California
Rio Conchos	21 January 1999	VI Rio Bravo
Rio San Juan (Bravo)	In process	VI Rio Bravo
Rio Turbio	9 February 1995	VIII Lerma-Santiago-Pacifico
Basin of the Chapala Lake	2 September 1998	VIII Lerma-Santiago-Pacifico
Ayuquila-Armeria	15 October 1998	VII Lerma-Santiago-Pacifico
Rio San Juan (Panuco)	1 August 1997	IX Golfo Norte

Source: Statistics of Water Resources in Mexico, CNA, 2003, Mexico (in Spanish).

Table 11.3: Basin committees as of 26 November 2002

Name	Date in which it was established	Administrative region
Rio Los Perros	18 November 1999	V Pacifico Sur
Rio Salado	18 May 2001	V Pacifico Sur
Rio Copalita	19 April 2002	V Pacifico Sur
Alto Atoyac	7 August 2002	V Pacifico Sur
Rio Blanco	16 June 2000	X Golfo Centro
Rio Zanatenco	23 August 2002	XI Frontera Sur
Canada de Madero	30 June 2000	IX Valle de Mexico

Source: Statistics of Water Resources in Mexico, CNA, 2003, Mexico (in Spanish).

The amended water law basically establishes that the thirteen administrative regions, as well as the state offices, will become basin organizations, but will depend on the Central government. This amended water law establishes that the basin organizations will be autonomous, but under the decision of the federal water authority, and also that the river basins are the units for water management in the country, but under the federal water authority. It remains to be seen how and by whom the decisions will be taken, and to what extent the river basin organisms will be autonomous. These are important considerations since it is likely that the federal government would continue to decide from the centre what is needed at the local levels. Interestingly, in the draft version of the water law, it was proposed the decentralization of the water authority from the Ministry of the Environment. However, this idea was rejected primarily because of financial considerations, since it would have meant that the Finance Ministry would have lost an annual revenue of approximately 7 billion dollars.

One of the main constraints for the water management of the country at the basin level is that the basin councils still remain a coordinating body, which can only make recommendations to the authorities and the users, but with no decision-making power. As stated in the Article 13 Bis 2 of the amended water law, 'The basin councils will be organized and will work based on what this (water) Law defines, as well as its regulation and the rules developed by the National Water Commission of Mexico.' A main constraint for river basin management stems from the fact that the councils are neither entitled to develop any regulations, nor can they execute any administrative or legal action. If the councils were operational, they could play a very important role in terms of planning as coordinating bodies.

Another important limitation in terms of management at the basin level is that this arrangement does not seem to be the most appropriate one for the implementation of the services at the municipal level, which follow a more traditional administrative structure. In addition, experiences have shown that a basin is more limited in the services it can provide to the users, compared to the municipalities (Sandoval, 2002), since decision making at the regional levels is still highly fragmented and the basins are more vulnerable than the states or municipalities (El Colegio de Mexico and CNA, 2003).

Cleary, one of the main problems is that the overall operational framework to manage water resources at the basin level is still not functional in the country: the main focus of water planning and management is still vested in one single institution at the central level, which has been unable thus far to respond adequately to the increasing needs of the sector. Even though twenty-five river basin councils have been established thus far, the federal government has still not formulated, let alone implemented, strategies to decentralize the functions, responsibilities, and funds, and transfer them to the basin councils and/or to the appropriate authorities at the regional, state or municipal levels. So far, only one out of the twenty-five river basin councils created is operational. In most cases, the councils do not even have staff or offices, not to mention implementable plans, financial support, and technical and management expertise. Even for this one operational case, no objective and independent assessment has even been made to evaluate its operational efficiency and impacts (Guerrero, 2004).

On the basis of their performances so far, the existing basin councils can neither be considered viable units for water management at the regional levels, nor can they become advisory organizations since they are subordinate to the interests of the central water authority. In fact, the basin councils can be considered within the country as virtual bodies. Fundamental legal changes will be necessary if they are to become successful institutions for regional water management in the future. In the case of the Lerma-Chapala basin (the most advanced basin council in the country), until very recently, most of the member states declined to participate in any meetings, arguing that that the legal framework did not give them any rights or responsibilities. At present, however, the representatives of the different sectors have realized that their participation is essential, if changes in the regions are to be achieved. Such attitudinal changes are likely to result in more participation, and in more autonomy for the stakeholders, who are gradually becoming actors instead of being spectators as was mostly the case earlier (Guerrero, 2000).

An institutional arrangement that has been put in place to reduce over-exploitation of aquifers is the bodies known as COTAS (Technical Committees for Groundwater according to CNA, and Technical Water Committees according to the Guanajuato State Water Commission). This is done through the development of new criteria for water allocation, with the participation of the several water users. Even though this is an innovative alternative, it is still not enough, since the COTAS are managed with fairly different approaches. According to the CNA, the COTAS should depend on the federal government, financially as well as in terms of authority, which naturally limits their actions and overall performances. However, the government of the state of Guanajuato has supported the COTAS as a true means for decentralization, both from institutional and financial viewpoints. The main constraint is that the Central authority still is not willing to give up its decision making role and economic power, and thus control, and the COTAS are not empowered with any legal authority, and thus are not likely to enforce any agreements or contribute to the reduction of the over-exploitation of the aquifers in the state (Marañón-Pimentel and Wester, 2000). It is worth noting that when Guanajuato developed its own master plan, which was formulated by the local experts, considering the local problems and proposing local alternatives, it faced very strong opposition from the central water authority throughout the process of plan formulation. On the basis of the performances thus far, it is evident that the central government is willing to transfer programmes, but not resources or authority (Guerrero, 1999 in Marañón-Pimentel and Wester, 2000).

To date, the characteristics of the COTAS being established by CNA remain unclear, specifically regarding their structures, tasks, and autonomy in relation to the federal government. It is likely, however, that the COTAS will not be financially and administratively autonomous organizations with an elected management board and a manager, as is the case of the water-users' associations in the transferred irrigation districts. Instead, they may be a forum in which aquifer users, government agencies' and the civil society will interact regarding groundwater management, under the auspices of the central water authority. In addition, they are organizations without legal status or decision making powers, whose recommendations may or may not be taken into account by the central authorities. On the other hand, the COTAS established in Guanajuato are responsible for surface and groundwater management, and only users are represented. Geohydrological studies for each aquifer, and mathematical models for most of them, have been developed, which allow the users to explore alternatives scenarios and plan their activities accordingly. By 2003, sixty two COTAS were established, out of which twenty had already become autonomous associations (Marañón-Pimentel, 2004).

Since the COTAS are a recent institutional innovation, it is still early to evaluate how effective they will be arresting the depletion of groundwater on a long-term basis (Wester et al., 2003).

Potential Application of Latin American Experiences to Asia

It is now fairly common to find in the area of natural resources management that the same paradigm is being proposed for use in all the countries of the world, irrespective of their physical, climatic, social, economic, legal, and institutional conditions, or the levels of development and management capacities available in the countries concerned. The field of water resources management is no exception to this approach. For example, two widely recommended paradigms for water management at present are IWRM and integrated river basin management (IRBM).

There is no question that the current approach to IRBM in the Latin American countries is influenced by the French experiences. Unfortunately, as of now, no serious, objective, independent, and in-depth study has been carried out as to how efficient the French model of IRBM has been, what are their advantages and shortcomings, and if the model has worked reasonably well or how sustainable it has been on a long-term basis. Only after such a study is properly completed, the next logical step would be to consider how appropriate it will be to transfer the French model to the Latin American countries. Since past experiences indicate that direct technology transfer between developed and developing countries are often fraught with danger, it is essential that studies are carried out to determine if the French model is the most suitable one, and, if so, what modifications are necessary so that this model could be used efficiently in specific Latin American countries.

On the basis of information available at present, it is difficult to conclude that IRBM will work in countries like Mexico, let alone recommending that the Mexican experiences could be applicable for the various Asian countries for many reasons. Among these reasons are the following:

(i)   The only river basin in Mexico where the integrated model has been tried to a certain extent is the Lerma-Chapala basin. No independent evaluation has been carried out of the results, both positive and negative, from the resulting experiences.

(ii) The legal and institutional structures in the Latin American countries owe more to the Spanish and the Portuguese systems, compared to the French system. There are some significant differences between these three approaches.

(iii) In a country like Mexico, water is under federal jurisdiction, and its management continues to be highly centralized and hierchial. The states have limited authority on water, and equally limited financial and human (management, technical, and administration) resources. Also, in spite of rethoric, decentralization of water management and public participation and involvement are only in their infancy.

(iv) One of the strengths of the French river-basin model is its power of taxation. This may not be possible under the Mexican political and institutional systems.

Even if the IRBM models work in the Latin American countries (there are serious concerns that it may not), its successful application in the Asian countries cannot be assumed to be a foregone conclusion. In an Asian country like India, water is under the jurisdiction of the states, and thus the federal government has limited control; its management is highly decentralized and the states have had financial and human resources to manage this resource for decades. In addition, the Indian legal system, institutions, and codes of practices owe more to the British system and not to the French, Spanish, or the Portuguese systems.

The world is heterogeneous. Countries have different cultures, social norms, physical attributes, and institutional arrangements. They face skewed availability of renewable and non-renewable resources, access to investment funds, and management and technical expertise. The systems of governance, legal frameworks, decision making processes, and types and effectiveness of institutions often differ from one country to another in many significant ways. Accordingly, under such widely diverse conditions, one fundamental question that needs to be asked is whether it is possible for one single paradigm of IRBM to be equally valid and universally applicable for all countries with diverse physical, economic, social, cultural, and legal conditions. Can this single paradigm be equally valid for an economic giant like the United States, technological powerhouse like Japan, world's most populous countries like China and India, and countries as diverse as Brazil, Bhutan, and Burkina Fasso? Can a single paradigm be equally applicable to Asian values, Latin American traditions, Japanese culture, Western and Islamic civilizations, and the emerging countries of Eastern Europe? These are difficult questions which the water profession has still not asked, let alone answered.

No conclusive statement can be drawn on the implications of the Latin American experiences in IRBM in terms of efficient water and related natural resources management, poverty alleviation or environmental conservation, based on very limited periods of data available, which have still not been independently assessed. What can be said with certainty is that given the types of constraints faced, it will be an extremely difficult task to apply IRBM successfully in most Latin American countries in the foreseeable future. Under these conditions, it is simply not possible to recommend that the Asian countries should consider IRBM based on the Latin American experiences available thus far.

Conclusions

Even though the current momentum in Latin America is moving towards the concept of IRBM, one cannot prejudge the results of this shift at the present stage. IRBM is a very complex subject, as are the institutional arrangements necessary to ensure their optimal functioning. When the results from these new management units are available, they need to be objectively, comprehensively, and seriously analysed. These results should be compared with other alternative institutional arrangements. Only after such a process, definitive conclusions can be drawn as to which should be the preferred option and under what conditions.

A major failure of the governments, not only in most Latin American countries but also in many other developing countries, has been that the river basin organizations have not yet been considered as real instruments for water resources management. It seems that their establishment has been more of a response to the current global trend, rather than the recognition of the actual benefits such institutional arrangements could bring to the management and planning of water resources at the regional and local levels. The river basin organizations are being created and they are being expected to function, but many times not even the support or the resources necessary are made available to them. Thus, their establishment seems to be an end by itself, and not the means to an end, the end being a more efficient, decentralized, participative process where needs and requirements of the several stakeholders (including the environment) are identified, analysed objectively, and then rational decisions are made within a social-economic-environmental framework.

Only two countries in the region, Brazil and Mexico, have developed legal and institutional frameworks for river basin organizations as units for water resources management. In other cases, where river basin organizations exist, they are not necessarily within the national policy-making framework, which makes their operation more complex and their results more difficult to predict. However, while the existence of legal and institutional frameworks within the national policy making is important and even essential, it does not assure either that the river basin organizations will be fully functional, that they would be supported by the administrative structures of the countries concerned, or that they would constitute the best option for efficient water management, until and unless the different interested parties commit on the long-term management of water, based on river basins as a unit for water management.

Present needs and future water requirements demand more flexible institutions and decision-making processes with emphasis on local and participatory structures. Implementation of decentralization has proved to be most difficult because of its complexities, and the reluctance of the central authorities to relinquish their control. This is mainly due to the fact that administrative and institutional structures in most of the countries have historically been vertical, where the paternalistic approaches still prevail and where decisions have traditionally been imposed from above, instead of being developed from below, as should be the case. Decision making would certainly be more dynamic and rational if it is approached and implemented on the basis of consultative, participatory, and transparent process, and not in terms of control.

The challenges facing the Latin American countries are enormous if the integration of all aspects of water resources is to be pursued at the river basin level. In the case of Mexico, the amended water law notes that the management of water resources is in a deep crisis which could be improved with new models of participation and a substantial change in the administration of water. However, only time will tell if the new institutional arrangements in terms of partial decentralization and continuous control by the central institutions will solve the present and future problems of the country through the implementation of the concept of integrated water resources management.

References

Barkin, D., and T. King, 1986, Desarrollo Economico Regional (Enfoque por Cuencas Hidrologicas de Mexico), Siglo XXI Editores, 5ª edición, Mexico.

Barrow, C., 'River Basin Development Planning and Management: A Critical Review', *World Development*, Vol. 26, No. 1, 1998, pp. 171–186.

Biswas, A.K. (ed.), *Water Resources of North America*, Berlin: Springer-Verlag, 2003.

Biswas, A.K., and C. Tortajada (eds), *Integrated River Basin Management, The Latin American Experience*, New Delhi: Oxford University Press, 2001.

Biswas, A.K., N. Cordeiro, B. Braga, and C. Tortajada, *Management of Latin American River Basins: Amazon, Plata and Sao Francisco*, Tokyo: United Nations University Press, 1999.

CNA, Statistics of Water in Mexico, National Water Commission of Mexico, Mexico, 2003 (In Spanish).

Dourojeanni, A., 'Challenges for Integrated Water Resources Management', Salvador: International Workshop on Water Policies and Institutions, 2000.

Dourojeanni, A., Experiences in establishing River Bsin Bodies in Latin America, XI Workshop of Water Law, Zaragoza, 2001a (In Spanish).

Dourojeanni, A., 'Public Policies for Integrated Watershed Management', in *Integrated River Basin Management for Latin America*, A.K. Biswas and C. Torajada (eds), New Delhi: Oxford University Press, 2001b.

ECLAC, Creation of Basin Organisms in Latin America and the Caribbean, Economic Commission for Latin America and the Caribbean, Santuiago, 1997 (In Spanish).

El Colegio de Mexico and CNA, Water for the Americas in the XXI Century, Colegio de Mexico, Mexico, 2003 (In Spanish).

Garcia, L., 'Experiences In the Preparation of Watershed Management Projects in Latin America', in *Integrated River Basin Management, The Latin American Experience*, Asit. K. Biswas and Cecilia Tortajada (eds), New Delhi: Oxford University Press, 2001.

Garcia, L., *Review of the Role of River Basin Organisations (RBO) in Latin America*, Washington D.C.: Inter-American Development Bank, 1999.

Global Water Partnership, Technical Advisory Committee, 'Integrated Water Resources Management, TAC Background Papers No. 4', GWP, Copenhagen 2000.

Guerrero-Reynoso, V., Proposal for the Decentralization of Water Management in Mexico by means of Basin Councils', Salvador: International Workshop on Water Policies and Institutions, 2000.

Guerrero, V., Contributions of Water Resources Management, in Towards a New Water Management in Mexico, C. Tortajada, V. Guerrero and R. Sandoval (eds.), Porrua, Mexico, 2004 (In Spanish).

Marañón-Pimentel, B., and P. Wester, Institutional Responses for the Management of Aquifers in teh Lerma-Chapala Basin, IWMI Latin American Series 17, Mexico, 2000 (In Spanish).

Marañón-Pimentel, B., Social Participation in the Management of Groundwater in Mexico, in Towards a New Water Management in Mexico, C. Tortajada, V. Guerrero and R. Sandoval (eds.), Porrua, Mexico, 2004 (In Spanish).

Meppem T., and S. Bourke, 'Different Ways of Knowing: A Communicative Turn Toward Sustainability', *Ecological Economics*, 30, 1999, pp. 389–403.

OECD, 'Territorial Reviews, Mexico', Mexico: Organization for Economic Cooperation and Development, 2003.

Sandoval Minero, R., Experiences in Guanajuato on Social Participation and Decentralization for Integrated Management of Water Resources based on River Basins, XI Workshop of Water Law, Zaragoza, 2001 (In Spanish).

SEMARNAT, Decree which modifies the Law of National Waters, Ministry of Environment and Natural Resources, DOF 29 April 2004, Mexico (In Spanish).

Solanes, M., and D. Getches, Laws and Regulations for Water Resources, ECLAC/Inter-American Development Bank, Report ENV-127, Washington, D.C., 1998.

Tortajada, C., 'River Basins: Institutional Framework and Management Options for Latin America', Cape Town: World Commission on Dams, Thematic Reviews, River Basin Institutional Framework and Management Options V3, 2000b.

Tortajada, C., *Sustainability of Water Resources Management in Mexico*, Mexico: Third World Centre for Water Management, 2000a.

Wester, P., R. Melville, and S. Ramos-Osorio, 'Institutional arrangements for water management in the Lerma-Chapala Basin', in *The Lerma-Chapala Watershed, Evaluation and Management*, Anne M. Hansen and Manfred van Afferden (eds), New York: Kluwer Academic/Plenum Publishers, 2003.

World Bank, 'Water Resources Management', A Bank Policy Paper, Washington D.C.: The World Bank, 1993, p. 140.

12

INTEGRATED WATER RESOURCES MANAGEMENT: A REASSESSMENT

Asit K. Biswas

Introduction

The need for water is universal. It is present everywhere. Without water, life, as we know it, will simply cease to exist. Water is constantly in motion, passing from one state to another, and from one location to another, which makes its rational planning and management a very complex and difficult task under the best of circumstances. Water may be everywhere, but its use has always been constrained in terms of availability, quantity, and quality.

Water problems of the world are neither homogenous, nor constant or consistent over time. They often vary very significantly from one region to another, even within a single country, from one season to another, and also from one year to another. Solutions to problems related to water depend not only on availability of water, but also on many other factors. Among these are the processes through which water is managed; the competence and capacities of the institutions that manage them; prevailing socio-political conditions which dictate water planning, development, and management processes and practices; appropriateness and implementation status of the existing legal framework; availability of investment funds; social and environmental conditions of the countries concerned; levels of available and usable technology; national, regional, and international perceptions; modes of governance including issues like political interferences, transparency, and corruption; educational and development conditions, and status, quality, and relevance of research that are being conducted on national, sub-national and local water problems.

Water is of direct interest to the entire population, as well as to most ministries of development at central and state levels, municipalities, private sector, and non-governmental organizations (NGOs). Such widespread interest is not an unique situation for water, as some water professionals have claimed: it is equally applicable to other issues like food, energy, environment, health, communication, or transportation. All these issues command high levels of attention in modern societies. In an increasingly interdependent and complex world, many issues are of pervasive interest for assuring good quality of life of the people. Water is one of these important issues, but it is certainly not the only important issue.

In recent years, it has become increasingly evident that the water problems of a country can no longer be resolved by water professionals, and/or the water ministries, alone. Problems pertaining to water are becoming increasingly more and more interconnected with other development-related issues, and also with social, economic, environmental, legal, and political factors, at local and national levels, and sometimes at regional and even international levels. Already, many of the problems have already become far too complex, interconnected, and large to be handled by any one single institution, irrespective of the authority and resources given to it, technical expertise and management capacity available, level of political support, and all the good intentions (Biswas, 2001).

The current and the foreseeable trends indicate that water problems of the future will continue to become increasingly more and more complex, and will become more and more intertwined with other development sectors like agriculture, energy, industry, transportation, and communication, and with social sectors like education, environment, and health and rural or regional development. The time is fast approaching when water can no longer be viewed in isolation by one institution, or any one group of professionals, without explicit and simultaneous consideration of other related sectors and issues, and vice versa. In fact, it can be successfully argued that the time has already come when water policies and major water-related issues should be assessed, analysed, reviewed, and resolved within an overall societal and development context. Otherwise the main objectives of water management, such as improved standard and quality of life of the people, poverty alleviation, regional and equitable income distribution, and environmental conservation, cannot be achieved. One of the main questions facing the water profession is how this challenge can be successfully answered in a socially-acceptable and economically efficient manner.

Integrated Water Resources Management

A few members of the water profession started to realize during the 1980s that the situation is not as good as it appeared. This feeling intensified during the 1990s, when many in the profession began to appreciate that the water problems have become multi-dimensional, multi-sectoral, and multi-regional, and filled with multi-interests, multi-agendas and multi-causes, and which can be resolved only through a proper multi-institutional and multi-stakeholder coordination. The issue at present, however, is not whether such a process is desirable, but rather how can this be achieved in the real world in a timely and a cost-effective manner.

Faced with such unprecedented complexities, many in the profession started to look for a new paradigm for management, which will solve the existing and the foreseeable problems related to water all over the world. The solution that was selected, was, however, not new. It was the rediscovery of a basically more than sixty-year-old concept, which could not be successfully applied earlier: integrated water resources management. Many who 'discovered' this concept were not aware that the 'new' concept was in fact was not really new, but has been around for several decades, but with a dubious record in terms of its implementation, which has never been objectively, comprehensively and critically assessed.

Before the status of application of integrated water resources management can be discussed to make water management more efficient, an important and fundamental issue that should first be considered is what precisely is meant by this concept. A comprehensive and objective assessment of the recent writings of individuals and institutions that are vigorously promoting integrated water resources management indicates that not only no one has a clear idea as to what exactly this concept means in operational terms, but also their views of it in terms of what it actually means and involves vary very widely.

The definition that is most often quoted at present is the one that was formulated by the Global Water Partnership (2000), which defined it as 'a process which promotes the coordinated development and management of water, land and related resources, in order to maximize the resultant economic and social welfare in an equitable manner without compromising the sustainability of vital ecosystems'.

This definition, on a first reading, appears broad, all-encompassing and impressive. However, such lofty phrases have little practical resonance on the present, or on the future water management practices. A serious and critical look may remind one of the immortal writings of William Shakespeare:

Polonius: What do you read, my lord?
Hamlet: Words, words, words.

The question that arises then is whether this well-intentioned and good-sounding definition has any real meaning in terms of its application and implementation to improve existing water management, or is it just an aggregation of trendy words collectively providing an amorphous definition which does not help water planners and managers very much in terms of actual application of the concept to solve real-life problems.

Let us consider some of the fundamental questions that the above definition raises in terms of its possible implementation in the real world, which have not been addressed to thus far:

- 'promotes': Who promotes this concept, why should it be promoted, and through what processes? Can the promotion of an amorphous concept be enough to improve water management? What about its implementation?
- 'land and related resources': What is meant by 'related resources'? Does it include energy, minerals, fish, other aquatic resources, forests, environment, etc.? In terms of land and agricultural resources, the water ministry mostly has no say or jurisdiction over them. Considering the intense inter-ministerial and intra-ministerial rivalries that have always been present in all countries, how can then use, development, and management of such resources be integrated, even if this was technically possible? Is this realistically feasible? If environmental and ecosystem resources are to be considered, how can water professionals and ministries handle such integration, which is often beyond their knowledge, expertise and/or control? Surprisingly, the people who formulated this definition for the Global Water Partnership are all from the water profession: experts from 'land and related resources' were singularly conspicuous by their absence. What makes the water profession believe that they can superimpose their views on the other professions, who were not even consulted? Equally, why should the professionals from other professions accept the view of some people from the water profession?
- 'maximize': What specific parameters are to be maximized? What process should be used to select these parameters properly? Who will select these parameters: only water experts as was the case for the formulation of the definition, or should experts from other areas be involved? What criteria should be used to select the necessary parameters? What methodology is available at present to maximize the selected parameters reliably? Do such methodologies even exist at present? If not, can they be developed within a reasonable timeframe?
- 'economic and social welfare': What exactly is involved in terms of determining economic and social welfare? Even economists and sociologists cannot agree as to what actually constitutes economic and

social welfare, except in somewhat general and broad terms, or how they can be quantified. Can water professionals 'maximize economic and social welfare' in operational terms, a fact that has mostly eluded the social scientists thus far? Is it possible to even establish cause-and effect relationships between water development and management and economic and social welfare, let alone be maximized? Such functional relationships are mostly unknown at present.

- 'equitable': What is precisely meant by equitable? How will this be determined operationally? Who decides what is equitable, for whom, and from what perspectives?
- 'sustainability': What is meant by sustainability, which itself is as vague a word, and also as fashionable, as integrated? How can sustainability be defined and measured in operational terms?
- 'vital ecosystems': What exactly constitutes vital ecosystems? How can 'vital' and 'non-vital' ecosystems be differentiated? Can even such a differentiation be made conceptually, let alone operationally? What are the minimum boundary conditions which will ensure the 'sustainability' of the 'vital ecosystems', irrespective of how sustainability itself is defined, or the issue to what constitutes vital ecosystems is resolved?

When all these uncertainties and unknowns are aggregated, the only objective conclusion that can be drawn is that even though on a first reading the definition formulated by the Global Water Partnership appears impressive, it really is unusable, or unimplementable, in operational terms. Not surprisingly, even though the rhetoric of integrated water resources management has been very strong in the various international forums during the past decade, its actual use (irrespective of what it means) has been minimal, even undiscernable in the field. In fact, one can successfully argue that it would not have made any difference in enhancing the efficiencies of macro-and meso-scale water policies, programmes, and projects of the recent years, even if the concept of integrated water resources management had not been rediscovered and promoted vigorously during the 1990s.

For all practical purposes, the definition that has been formulated by the Global Water Partnership is unimplementable. In addition, it is internally inconsistent. Furthermore, while it uses many of the current trendy words, it does not provide any real guidance to the water professionals as to how the concept can be used to make the existing water planning, management and decision-making processes increasingly more and more rational, efficient and equitable.

The definition of integrated water resources management is an important consideration. When the definitional problem can be successfully resolved in an operational manner, it may be possible to translate it into measurable criteria, which can then be used to appraise the degree to which the concept of integration has been applied in a specific case, and also the overall relevance and usefulness of the concept.

In addition, a fundamental question that has never been asked, let alone answered, or for which there is no clear-cut answer at the present state of knowledge, is what are the parameters that need to be monitored to indicate that a water resources system is functioning in an integrated manner, or a transition is about to occur from an integrated to an unintegrated stage, or vice versa, or indeed even such a transition is occurring. In the absence of both an operational definition and measurable criteria, it is not possible to identify what constitutes an integrated water resources management at present, or how should water be managed so that the system remains inherently integrated on a long-term basis.

There is no question that in the area of water resources, integrated water resources management has become a powerful and all-embracing slogan during the past ten years. This is in spite of the fact that operationally it has not been possible to identify a water management process which can be planned and implemented in such a way that it becomes inherently integrated, however this may be defined, right from its initial planning stage and then to implementation and operational phases. For all practical purposes, most international institutions have endorsed this concept, either explicitly or implicitly. This is despite the fact that there is no agreement at present among the same international institutions that endorse it as to what exactly is meant by integrated water resources management, or whether it has been possible to use this concept to improve water management practices which would not have occurred under normal circumstances, and without any explicit use of this concept. Also lacking is a consensus regarding the countries where it has been possible to apply this concept successfully, and, if so, under what conditions, over what periods, and what have been its impacts (positive, negative, and neutral) on human lives, environment, and other appropriate development indicators, and how, or even whether, the concept can be implemented in the real world.

This type of almost universal popularity of a vague, undefinable, and unimplementable concept is not new in the area of resources management. It has happened many times in the past. During the twentieth century, many such popular concepts have come and gone, without leaving much of a footprint on how natural resources can be managed on a long-term basis. Such concepts generally became politically correct during the time of their popularity, and are vague enough for everyone to jump on the bandwagon and claim they are following the latest paradigm. In fact, it can be argued that the vagueness of a concept to a significant extent increases its popularity, since people can easily continue to do whatever they were doing before, but at the same time claim that they are following the latest paradigm.

The current popularity of the concept reminds one of another similar concept which received wide popular support in the United States during the early twentieth century: conservation. Even President Roosevelt of the United States said at that time: 'Everyone is for conservation: no matter what it means!' (Biswas, 2001). The situation is very similar in the early part of the twenty-first century with integrated water resources management. To paraphrase, and perhaps update President Roosevelt, one can say that 'everyone is for integrated water resources management: no matter what is means, no matter whether it can be implemented, or no matter whether it would actually improve water management processes'. The only difference between the conservation movement of President Roosevelt's time and the movement on integrated water resources management of the present is that the information and communication revolution and the globalization processes have ensured that the gospel of integrated water resources management has been spread all over the world, and not mostly confined to the United States, as was mostly the case for the conservation movement earlier.

The concept of integrated water resources management was promptly embraced by many international institutions during the 1990s, many of whom were not even aware that it has been around for more than half a century. Thus, and perhaps not surprisingly, the authors of Toolbox for IWRM for the Global Water Partnership claimed erroneously in 2003 that 'IWRM draws its inspiration from the Dublin principles', being blissfully unaware of the longevity of this concept, or that international institutions like the United Nations were promoting this concept, extensively during the 1950s, or the United Nations Water Conference, held in Mar del Plata, Argentina, in March 1977, had more to say on IWRM (Biswas, 1978) than the so-called Dublin Conference. In addition, the Mar del Plata Conference was an intergovernmental meeting, and its Action Plan (which included integrated water resources management) was endorsed by all the governments that were members of the United Nations in 1977. In contrast, the Dublin Conference of 1992 was a meeting of experts, and thus its recommendations, whatever they were, were never approved by any government, irrespective of the claims to the contrary of the individuals and institutions that were mostly responsible for the organization of this conference.

Extensive and intensive analyses of literature on integrated water resources management published during the past decade indicate two somewhat unwelcome developments. First, there is no clear understanding of what exactly integrated water resources management means. Accordingly, different people have interpreted this concept very differently, but under a very general catch-all concept of integrated water resources management. Absence of any usable and implementable definition has only compounded the vagueness of the concept, and has reduced its implementation potential to a minimum. Second, because of the current popularity of the concept, some people have continued to do what they were doing in the past, but under the currently fashionable label of integrated water resources management in order to attract additional funds, or to obtain greater national and international acceptance and visibility.

An analysis of even the issues that should be integrated under the IWRM level, indicates a very wide divergence of opinions. It should be noted that this refers only to what should be integrated, and not to other equally important fundamental issues like how these issues can be integrated (even they can actually be integrated since many of the issues are mutually exclusive), who will do the integration and why, what processes will be used for integration (do such processes currently exist?), or will the integration, if at all it can be done, produce the benefits that proponents now claim. Regrettably, none of these questions are now being even seriously asked, let alone having objective and definitive answers.

Analyses of existing literature indicate that the authors concerned have considered different issues that need to be integrated under this concept. Thus, depending upon the author(s) concerned, integrated water resources management means integration of:

- objectives which are not mutually exclusive (economic efficiency, regional income redistribution, environmental quality and social welfare);
- water supply and water demand;
- surface water and groundwater;
- water quantity and water quality;
- water and land-related issues;
- different types of water uses: domestic, industrial, agricultural, navigational, recreational, environmental, and hydropower generation;

- rivers, aquifers, estuaries, and coastal waters;
- water, environment, and ecosystems;
- water supply and wastewater collection, treatment, and disposal;
- macro, meso, and micro water projects and programmes;
- urban and rural water issues;
- water-related institutions at national, regional, municipal, and local levels;
- public and private sectors;
- government and NGOs;
- timing of water release from the reservoirs to meet domestic, industrial, agricultural, navigational, environmental, and hydropower generation needs;
- all legal and regulatory frameworks relating to water, not only directly from the water sector, but also from other sectors that have implications on the water sector;
- all economic instruments that can be used for water management;
- upstream and downstream issues and interests;
- interests of all different stakeholders;
- national, regional, and international issues;
- water projects, programmes, and policies;
- policies of all different sectors that have implications for water, both in terms of quantity and quality, and also direct and indirect (sectors include agriculture, industry, energy, transportation, health, environment, education, gender, etc.);
- intra-state, interstate, and international rivers;
- bottom-up and top-down approaches;
- centralization and decentralization;
- national, state, and municipal water policies;
- national and international water policies;
- climatic, physical, biological, human, and environmental impacts;
- all social groups, rich and poor;
- beneficiaries of the projects and those who pay the costs;
- service providers and beneficiaries;
- present and future generations;
- national needs and interests of donors;
- activities and interests of donors;
- all gender-related issues;
- present and future technologies; and
- water development and regional development.

The above list, which is by no means comprehensive, identifies 37 sets of issues which different authors consider to be the issues that should be integrated under the aegis of integrated water resources management. Even at a conceptual level, all these 37 sets of issues that the proponents would like to be integrated, simply cannot be achieved.

These types of fundamental issues need to be discussed and resolved successfully before the concept of integrated water resources management can be holistically conceived, and then serious efforts can be made to implement it. Unfortunately, while much lip service is given to this concept at present, most of the published works on the subject are somewhat general, or a continuation of earlier 'business as usual' undertakings, but with a trendier label of integrated water resources management. If integrated water resources management is to become a reality, national and international organizations will have to address many real and complex questions, which they have not done so far in any meaningful fashion, nor are there any indications that they are likely to do so in the foreseeable future. Under these circumstances, and unless the current rhetoric can be translated effectively into operational reality, integrated water resources management will remain a fashionable and trendy concept for some years, and then gradually fade away like many other similarly popular concepts of earlier times.

Popularity of the Concept

An important issue that needs to be asked is why an old concept suddenly became so popular in the 1990s, to the extent that some people and institutions now consider it to be the 'holy grail' of water management? There are many reasons for its sudden leap of popularity, and only some of the main reasons will be discussed herein.

Probably the most important reason for its current popularity is the attractiveness of the concept at least at a superficial level. In a world that operates on the principle of reductionism, integrated water resources management gives a feeling of using a comprehensive and holistic approach, which many people a priori assume will produce the best results irrespective of its shortcomings and certain inherent inconsistencies. The time has come to review this aspect objectively.

Historically, it was possible for a brilliant person to know nearly all there was to know until about the end of the sixteenth century. Versatile geniuses like Aristotle, Theophrastus, Vitruvias, Isidore of Seville, and Leonardo da Vinci could discuss most subjects authoritatively. Human knowledge of natural and social sciences were at a stage where it was possible for a truly gifted person to master all the knowledge that was available during their lifetimes.

The situation started to change around the seventeenth century. By the early eighteenth century, tremendous advances in knowledge had made it impossible for anyone to be a universal encyclopaedic, and keep up with the constant generation of new knowledge. This realization was gradually reflected in the development of a new branch of knowledge, which initially became known as natural philosophy, and began to be distinguished increasingly from traditional philosophy, which was earlier considered to be the exclusive discipline for knowledge. The nineteenth century witnessed exponential advances in human knowledge and, with it, technological developments. It was no longer possible for any one individual to master natural philosophy completely. Thus, new disciplines began to emerge, which further fragmented the knowledge base to manageable levels. Natural philosophy was subsequently subdivided, initially into physics and chemistry, and later further to other disciplines like life sciences and biological sciences.

The information explosion of the twentieth century accelerated this trend towards reductionism. Disciplines became more and more fragmented. It became humanly impossible for anyone to know as much there is to know even on a much more restricted subject area like water. Knowledge, communication, and information revolution, and increasing globalization witnessed towards the end of the twentieth century, further restricted one's disciplinary knowledge base. With the frontiers of knowledge expanding continuously, it is becoming increasingly difficult for professionals to keep up with the advances even in their limited areas of interest.

As the world became increasingly more and more complex, the disciplinary knowledge base of individuals started to reduce as well. People started to specializing in narrower and narrower subject areas. Concomitantly, managing human societies became more and more complex, as a result of which new institutional machineries had to be created with increasingly narrower focuses. New institutions had to be created in areas which were part of broader groups earlier. For example, in 1972, when the United Nations Conference on the Human Environment was held in Stockholm, only eleven countries had environmental machineries. Two decades later, nearly all countries of the world had similar institutions. For a variety of reasons, including efficient management, smaller institutions were preferred compared to humongous ones.

Thus, in recent centuries, a progressively reductionist approach has been applied to both knowledge and institutions. Integrated water resources management, in a sense, can be viewed as a nostalgic approach to a broader and more holistic way to manage water, as may have been possible in the past. However, since the world has moved on, water management needs to move with it.

In one sense, integrated water resources management, irrespective of the general impression prevalent in the water profession, is not holistic. This is not surprising, since most water professionals consider, at least implicitly, water to be very important, if not the most important resource. The other issues like energy, agriculture or environment do not generally receive appropriate emphasis or consideration, though some receive comparatively more attention than others.

If integrated water resources management is considered essential by the water profession, other disciplines can justifiably promote similar concepts like integrated energy resources management, or integrated agricultural management, or integrated environmental management. Unfortunately, in a complex world, issues like water, energy, agriculture, or environment are becoming increasingly interrelated and interdependent, and thus integrated management of any one of these resources is not possible because of accelerating overlaps and interlinkages with the other resources. Developments in the water area invariably affect management of resources like energy, agriculture, or ecosystems, and the developments in these resource areas, in turn, affect water.

Let us consider the issue of water and energy interrelationships. The water profession has mostly ignored energy, even though in many ways water and energy are closely interlinked. For example, water not only produces energy (hydropower), but the water sector is also a prodigious user of energy. Accordingly, in a country like India, hydropower accounts for slightly over 20 per cent of electricity generated, but the water sector in turn 'consumes' similar amount of India's electricity. Furthermore, no large-scale electricity production, be it thermal, nuclear or hydro, is possible without water. In countries like France, the biggest user of water is not agriculture, but the energy industry. Thus, it simply is not possible to consider water resources management in an integrative manner without reference to energy, or integrated energy resources management without considering water. In other words, both technically and conceptually, it is not possible to consider parallel efforts which will focus exclusively on integrated management of water or energy because of their inherently extensive and intensive overlaps and interlinkages.

Since water and energy are interrelated, consideration of integrated water resources management per se could contribute to unintegrated energy management, since these two resources have many common factors in terms of management. Both these two resources cannot be separately planned in an 'integrative' manner, irrespective of how integration is defined. Optimizing the benefits of integrated water resources management, even if this can be operationally achieved by a miracle, will not result in the maximization of the benefits of integrated energy management, or vice versa. There will be trade-offs, both positive and negative, for any such management approaches for these two resources in an independently integrated manner.

One can conceivably argue that if water and energy cannot be managed in an integrative manner independently, perhaps these two resources can be managed together as integrated water and energy resources management. This is also not a practical solution because while there are significant interlinkages between water and energy, the processes available at present for their overall management are very different, and the expertise needed to manage these two resources efficiently are also very different. Furthermore, institutionally, if these two resources are combined under one umbrella, for most countries it will result in a large and unmanageable institution, which is likely to be both undesirable and counterproductive. In a few countries, at least institutionally, water and energy are managed by the same governmental ministry. These countries, however, are mostly comparatively small, and thus the management of these two resources by one institution is still feasible. This, however, is not possible for large countries like Brazil or India.

If the current global institutional arrangements for management of water and energy resources are analysed, they are often somewhat arbitrary. For example, hydropower in some countries is placed within the mandate of the Ministry of Electricity or Energy, which means that the Ministry of Water has very limited say as to how hydropower projects are planned, operated and managed. In some other countries, the Water Ministry is responsible for hydropower, even though hydropower contributes to a very significant percentage of national electricity generation. There is thus no simple and elegant solution. It is also interesting to note that in a country like Canada, the word 'hydro' is synonymous with electricity, even though water and electricity are managed very differently, both technically and institutionally.

Irrespective of whether hydropower is located institutionally within the Ministry of Energy or Water, it is likely to contribute to the non-optimal integration of the management of these two resources. What is thus needed is not integration in terms of management of these two resources, but close collaboration, cooperation and coordination between the two institutions, as well as other public and private sector institutions associated with their management. In the real world, such collaborations are unfortunately limited, and often somewhat ad-hoc. They also vary with time, even for the same country. One is reminded of Voltaire's assertion that 'best is the enemy of good'. The 'best' solutions for integrated water management and integrated energy management may not be compatible. What we can strive for is a 'good' solution which could result in acceptable management practices for both water and energy.

The problem becomes even more complex since it is not only the energy sector that is closely linked to water, but also other sectors like agriculture, environment or industry. Globally, agricultural sector is the largest user of water. Thus, neither agriculture nor water can be managed in an 'integrated' way without considering the other. The issue becomes even more unmanageable if parallel efforts are made to manage water, energy, agriculture, industry, and/or environmental sectors in an integrated manner, however the word integrated be defined. Thus, integrated water resources management at a first and somewhat superficial view may appear to be a holistic approach, but on deeper consideration, it ends up as a reductionist approach, but perhaps at a somewhat higher level.

Accordingly, integrated management of a specific resource like water cannot simply be considered to be a holistic approach. One can argue that it may be possible to manage two or more natural resources by combining their management processes. Past experiences indicate that this is generally neither a practical nor efficient solution. A good example is what happened in Egypt, during the 1970s, when the Ministries of Irrigation and Agriculture were combined, so that this combined entity would manage these two sectors more rationally and efficiently than what was in the past. The then Minister of Irrigation, who probably was one of the most dynamic and competent Minister of Irrigation that Egypt has had since President Nasser's revolution in 1952, became the minister of this new enlarged institution. In spite of his heroic and strenuous efforts, it was simply not possible to manage the new ministry efficiently or integratively. After a very short period, the management process was reversed: irrigation and agriculture became two separate ministries again. This practice has continued ever since, even though the name of the irrigation ministry was changed twice subsequently. In spite of the name changes, this ministry has basically remained a water management institution, like in the vast majority of the other countries of the world.

In the real world, integrated water resources management, even in a limited sense, becomes difficult to achieve because of extensive turf wars, bureaucratic infighting, and legal regimes (like national constitutions), even within the management process of a single resource like water, let alone in any combined institution covering two or more ministries which have been historic rivals. In addition, merger of such institutions produce a humongous organization that is neither easy to manage nor control.

It should also be noted that water has linkages to all developmental sectors and social issues like poverty alleviation and regional income distribution. It is simply unthinkable and totally impractical to bring them under one roof in the guise of integration, irrespective of how it is defined. Such integrations are most likely to compound the complexities of the problems, instead of solving them.

Some have argued that integrated water resources management is a journey, and not a destination, and the concept provides only a road map for the journey. The problems, however, with such reasoning is that in the area of water management, we are long on road maps, but short on drivers! Equally, road maps may be useful, but in order to use them we need a starting point and a destination. Without knowing the starting point and the destination, road maps are of very limited use since one is most likely to be all over the place. Another problem of using a road map analogy for integrated water resources management is that we do not know where we wish to go, except in a very vague manner, and since we have no idea as to how to identify the final destination, we would have no idea when we have reached that destination, even if we reach the destination by a fluke. Not knowing the destination, it is not possible to decide if we are travelling in the right direction, or the probability of reaching the right end. In the final analysis, it is not very helpful to be long on concepts but short on their implementation potential.

There could be also some negative implications of integrated water resources management which for the most part have not been seriously considered.

Already, in a few countries, there are indications that the main national water institution is trying to take over other water-related institutions in the name of better integration. The implicit assumption is such consolidation of institutions will contribute to integrated water resources management. However, this may not be an efficient approach since different institutions have different stakeholders and interests, and this diversity is a part of any democratic process. The consolidation of institutions, in the name of integration, is likely to produce more centralization, and reduced responsiveness of such institutions to the needs of the different stakeholders, which is not an objective that the present societies and international institutions prefer at present. Water management must be responsive to the needs and demands of a growing diversity of central, state and municipal institutions, user groups, private sector, NGOs, and other appropriate bodies. Concentration of authority into one or fewer institutions could increase biases, reduce transparency and proper scrutiny of their activities.

In addition, objectives like increased stakeholder participation, decentralization and decision-making at the lowest possible level are unlikely to promote integration, however, it is defined, under most conditions, especially for meso and macro water projects.

Integrated water resources management, like other similar concepts (e.g., integrated rural development, or integrated area development) have historically run into very serious difficulties in terms of their implementation. Conceptually they could be considered to be attractive paradigms, but the world is complex, and many concepts, irrespective of their initial attractiveness and simplicity, cannot be applied to solve increasingly complex and interdependent issues and activities (Biswas and Tortajada, 2004). Even after more that half a century of existence, it has not been possible to find a practical framework that could be used for the integration of the various issues associated with water management.

Conclusions

It is argued that integrated water resources management has become a popular concept in recent years, but its application to more efficiently manage macro- and meso-scale water policies, programmes, and projects, has been dismal. Conceptual attraction by itself is not enough: concepts, if they are to have any validity, must be implementable to find better and more efficient solutions. This is not only not happening at present, but there are also no signs that the situation is likely to change in the foreseeable future.

It is also necessary to ask a very fundamental question: why it has not been possible to properly implement a concept that has been around for some two generations in the real world for macro- and meso-level water projects and programmes? The question then arises is that is the concept of integrated water resources management is a universal solution as its many proponents currently claim, or is it a concept that has limited implementation potential, irrespective of its conceptual attractiveness and current popularity? Unless the concept of integrated water resources management can actually be applied in the real world to demonstrably improve the existing water management practices, its current popularity and extensive endorsements by international institutions become irrelevant. Knowledge, fortunately, does not advance by consensus: if it did, we would still be living in the Dark Ages!

In addition, the world is heterogeneous, with different cultures, social norms, physical attributes, skewed availability of renewable and non-renewable resources, investment funds, management capacities, and institutional arrangements. The systems of governance, legal frameworks, decision-making processes, and types and effectiveness of institutions often differ from one country to another in very significant ways. Accordingly, and under such diverse conditions, one fundamental question that needs to be asked is whether it is possible for a single paradigm of integrated water resources management to encompass all countries, or even regions, with diverse physical, economic, social, cultural, and legal conditions? Can a single paradigm of integrated water resources management be equally valid for an economic giant like the United States, technological powerhouse like Japan, and for countries with diverse conditions as Brazil, Bhutan, or Burkina Faso? Can a single concept be equally applicable for Asian values, African traditions, Japanese culture, Western civilization, Islamic customs, and emerging economies of Eastern Europe? Can any general paradigm be equally valid for monsoon and non-monsoon countries, deserts and very wet regions, and countries in tropical, sub-tropical, and temperate regions, with very different climate, institutional, legal, and environmental regimes? The answer most probably is likely to be no.

What is now needed is an objective, impartial, and undogmatic assessment of the applicability of integrated water resources management. Unfortunately, most of its current promoters have a priori assumed that this concept will automatically make the water management processes and practices ideal. Equally, unfortunately, current evidences indicate that irrespective of the current popularity of the concept, its impact to improve water management has been, at best, marginal. A cynic might even say that we sit in watertight compartments, but preach holistic approaches to water management. Perhaps, the salutary caution of Harold Macmillan, the former Prime Minister of the United Kingdom, is appropriate in the current context: 'After a long life I have come to the conclusion that when all the establishment is united, it is always wrong!' Is it possible that this cautionary statement is applicable to integrated water resources management as well?

References

Biswas, Asit K., 'Water Policies in the Developing World', *International Journal of Water Resources Development*, Vol. 17, No. 4, 2001, pp. 489–99.

Biswas, Asit K. (ed.), 'United Nations Water Conference: Summary and Main Documents', Oxford: Pergamon Press, 1978, p. 217.

Biswas, Asit K., and Cecilia Tortajada (eds), *Appraising the Concept of Sustainable Development: Water Management and Related Environmental Challenges*, New Delhi: Oxford University Press, 2004.

Global Water Partnership, 'Integrated Water Resources Management Toolbox, Version 2', Stockholm: GWP Secretariat, 2003, p. 2.

Global Water Partnership, 'Integrated Water Resources Management', TAC Background Papers No. 4, Stockholm: GWP Secretariat, 2000, p. 22.

Tortajada, Cecilia, Olli Varis and Asit K. Biswas (eds), 'Integrated Water Resources Management in South and Southeast Asia', Oxford University Press, 2004.

INDEX

344   *Index*